Foundations of Convex Geometry

AUSTRALIAN MATHEMATICAL SOCIETY LECTURE SERIES

1 Introduction to Linear and Convex Programming, N. CAMERON
2 Manifolds and Mechanics, A. JONES, A. GRAY & R. HUTTON
3 Introduction to the Analysis of Metric Spaces, J. R. GILES
4 An Introduction to Mathematical Physiology and Biology, J. MAZUMDAR
5 2-Knots and their Groups, J. HILLMAN
6 The Mathematics of Projectiles in Sport, N. DE MESTRE
7 The Peterson Graph, D. A. HOLTON & J. SHEEHAN
8 Low Rank Representations and Graphs for Sporadic Groups, C. PRAEGER & L. SOICHER
9 Algebraic Groups and Lie Groups, G. LEHRER (ed)
10 Modelling with Differential and Difference Equations, G. FULFORD, P. FORRESTER & A. JONES
11 Geometric Analysis and Lie Theory in Mathematics and Physics, A. CAREY & M. MURRAY (eds)

Foundations of Convex Geometry

W. A. COPPEL
Department of Theoretical Physics,
Australian National University

CAMBRIDGE
UNIVERSITY PRESS

CAMBRIDGE UNIVERSITY PRESS
Cambridge, New York, Melbourne, Madrid, Cape Town, Singapore, São Paulo

Cambridge University Press
The Edinburgh Building, Cambridge CB2 8RU, UK

Published in the United States of America by Cambridge University Press, New York

www.cambridge.org
Information on this title: www.cambridge.org/9780521639705

First published 1998

A catalogue record for this publication is available from the British Library

Library of Congress Cataloguing in Publication data

Coppel, W. A.
Foundations of convex geometry / W. A. Coppel.
p. cm. – (Australian Mathematical Society lecture series: 12)
Includes bibliographical references and index.
ISBN 0 521 63970 0
1. Convex geometry. I. Title. II. Series.
QA639.5C66 1998
516'.08 – dc21 97–32149 CIP

ISBN 978-0-521-63970-5 paperback

Transferred to digital printing 2007

For Maria Antonietta

Contents

Preface

Euclid's *Elements* held sway in the mathematical world for more than two thousand years. (For an English translation, see Euclid (1956).) In the nineteenth century, however, the growing demand for rigour led to a re-examination of the Euclidean edifice and the realization that there were cracks in it. The critical reappraisal which followed was synthesized by Hilbert (1899), in his *Grundlagen der Geometrie*, which has in turn held sway for almost a century.

In what ways may Hilbert's treatment be improved upon today? Apart from technical improvements, a number of which were included in later editions of his book, it may be argued that Hilbert followed Euclid too closely. The number of undefined concepts is unnecessarily large and the axioms of congruence sit uneasily with the other axioms. In addition, the restriction to three-dimensional space conceals the generality of the results and seems artificial now.

For these reasons Hilbert's approach is often replaced today by a purely algebraic one – the axioms for a vector space over an arbitrary field, followed by the axioms for a vector space over the real field with a positive definite scalar product. This permits a rapid development, but it simply begs the question of why it is possible to introduce coordinates on a line. Only when one attempts to answer this question does one realise that many important results in no way depend on the introduction of coordinates. It is curious that some powerful advocates of 'coordinate-free' linear algebra have used real analysis to prove purely geometric results, such as the Hahn–Banach theorem, without any apparent twinge of conscience.

A system of axioms for Euclidean geometry in which the only undefined concepts are *point* and *segment* was already given before Hilbert by Peano. We follow his example in the present work, although some of our axioms differ. The choice of a system of axioms is inherently arbitrary, since there will be many equivalent systems. However, the purpose of an axiom system is not only to provide a basis for rigorous proof, but also to reveal the structure of a subject.

From this point of view one axiom system may seem preferable to another which is equivalent to it. The development should seem natural, almost inevitable.

Considerations of this nature have led us to isolate a basic structure, here called a *convex geometry*, which is defined by two axioms only. Additional axioms are chosen so that each individually, in conjunction with these, guarantees some important property. Four such additional axioms define another basic structure, here called a *linear geometry*, for which a rich theory may be developed. Examples of linear geometries, in addition to Euclidean space, are hyperbolic space and (hemi)spherical space, where a 'segment' is the geodesic arc joining two points.

Apart from a dimensionality axiom, only seven axioms in all are needed to characterize ordinary Euclidean or, more strictly, *real affine* space. However, dimensionality plays a role in achieving this number, by eliminating some other possibilities in one or two dimensions. Although the characterization of Euclidean space may be regarded as our ultimate goal, it would be contrary to our purpose to impose all the axioms from the outset. Instead we adjoin axioms successively, so that results are proved under minimal hypotheses and may be applied in other situations, which are of interest in their own right. For example, Proposition III.17 establishes the theorems of Helly and Radon in any linear geometry, and Chapter IV similarly extends the facial theory of polytopes. In this way each result appears, so to speak, in its 'proper place'. When there is a branching of paths, we choose the one which leads to our ultimate goal. This approach, I believe, has not previously been pursued in such a systematic manner. I have found it illuminating myself and hope that the illumination succeeds in shining through the present account.

We have spoken of the characterization of Euclidean space as our ultimate goal, but we are actually concerned with characterizing only a convex subset of Euclidean space. The greater freedom of the whole Euclidean space is mathematically desirable, but it is reasonable to require only that our physical world may be embedded in such a space. This type of non-Euclidean geometry was already mentioned by Klein (1873) and was further considered by Schur (1909). In view of the great historical importance of the parallel axiom, it is of interest that it can play no role here, since it fails to hold in any proper convex subset of Euclidean space (except a lower-dimensional Euclidean space).

The foundations of geometry have been studied for thousands of years, and thousands of papers have been written on the subject. Since it is impossible to do justice to all previous contributions in such a situation, we have deliberately restricted the number of references. Thus the absence of a reference does not necessarily imply ignorance of its existence and is not a judgement on its quality. Some references are included simply to provide a time scale and others for their useful bibliographies. Our task has been to select and organize, and sometimes extend. Some open problems are mentioned at the ends of Chapters III, IV and VIII.

The topic of axiomatic convexity has been omitted from the recent extremely valuable *Handbook of convex geometry* (Vols. A and B, ed. P.M. Gruber and J.M. Wills, North-Holland, Amsterdam, 1993), and ordered geometry receives only passing reference (on p. 1311) in the equally valuable *Handbook of incidence geometry* (ed. F. Buekenhout, North-Holland, Amsterdam, 1995). It is hoped that the present work may in some measure repair these omissions. Our aim has been to give a connected account of these subjects, which may be read without reference to other sources. Since many results are established under weaker hypotheses than usual, rather detailed proofs have been given in some cases.

The work is not arranged as a textbook, with starred sections and exercises, and is perhaps more difficult than the usual final-year undergraduate or first-year graduate course. However, the mathematical prerequisites are no greater and I believe that an interesting course could be constructed from the material here, which would acquaint students with a cross-section of mathematics in contrast to the usual compartmentalized course.

Familiarity is assumed with the usual language and notation of set theory, with such algebraic concepts as group, field and vector space, and with Dedekind's construction of the real numbers from the rationals. Concepts from other areas, such as partially ordered sets and lattices, projective geometry, topology and metric spaces, are defined in the text. A few elementary properties of metric spaces are stated without proof. On the two or three occasions when appeal is made to some other result the subsequent development is not at stake. Although the work is essentially self-contained, the reader is still encouraged to consult other treatments, both in the references cited in the notes at the end of each chapter and in the introductory chapters of more specialized works on convexity theory, such as Bonnesen and Fenchel (1934), Valentine (1964), Leichtweiss (1980) and Schneider (1993). The remaining prerequisites for the present work are more substantial – the ability and resolve to follow a detailed logical argument and a love of mathematics.

For assistance in various ways I thank M. Albert, B. Davey, V. Klee, J. Reay, J. Schäffer, V. Soltan and H. Tverberg. I am especially grateful to H. Tverberg for the detection of errors, misprints and obscurities in the original manuscript. The exposition and index have been improved by suggestions from the referees, and the appearance by suggestions from the copy-editor.

As I write these lines on the eve of my retirement from paid employment in the Institute of Advanced Studies at the Australian National University, I take the opportunity to acknowledge that this book could not have been written without the privileged working conditions which I have enjoyed.

Andrew Coppel

Introduction

In the present work convex subsets of a real vector space are characterized by a small number of axioms involving *points* and *segments*. At the same time a substantial part of the basic theory of convex sets is developed in a purely geometric manner. This contrasts with traditional treatments, in which the origin has a distinguished role and in which use is made of such analytic devices as duality and the difference of two sets, not to mention metric properties.

We owe to Pasch (1882) the recognition of the role of *order* in Euclidean geometry, both the order of points on a line and a more subtle order associated with points in a plane. A basic undefined concept in Pasch's work was the segment determined by two distinct points. By assuming various axioms involving segments he was able to define a *line* and the order of points on a line. An additional undefined concept was that of the *plane* determined by three points not belonging to the same line.

Peano (1889) avoided this additional undefined concept by replacing Pasch's planar ordering axiom by two further axioms involving segments. Thus he was able to give a system of axioms for Euclidean geometry in which the only undefined concepts were point and segment. Some of Peano's axioms were consequences of the remaining axioms. An independent system of axioms, involving only points and segments, was given by Veblen (1904).

Two axioms which were used from the outset by all these authors will be mentioned here. The first, called *density* in the present work, states that the segment $[a,b]$ determined by two distinct points a,b contains a point $c \neq a,b$. The second, called *unendingness* in the present work, states that for any two distinct points a,b there is a point $c \neq a,b$ such that b is contained in the segment $[a,c]$ determined by a,c. In recent years discrete mathematics has attained equal status with continuous mathematics, and it seems desirable to develop the foundations of geometry as far as possible without using these two axioms. This is what is done here.

One consequence is that our segments are closed (endpoints included), rather than open (endpoints excluded) as in the work of the authors cited. However, this is not a novelty; see, for example, Szczerba and Tarski (1979). What is new here is the gradual introduction of axioms and an obstinate insistence on establishing results without the use of axioms which are not required.

I
Alignments

This chapter is of a preparatory nature. For convenience of reference we begin by defining some basic concepts related to partially ordered sets and lattices. We also prove Hausdorff's maximality theorem, which will be used repeatedly. However, the main topic of the chapter is alignments. An *alignment*, like a topology, is a collection of subsets with certain properties. (The name is not very suggestive, but has the merit that it is not used elsewhere in mathematics.) Alignments, or equivalently *algebraic hull operators*, provide an abstract framework for such concepts as *extreme point, independent set, basis* and *face*. *Helly sets, Radon sets* and *Carathéodory sets* are also defined. Although they are of interest in the present general context, they will acquire a more familiar form in Chapter III.

Two particular types of alignment are given special attention, those with the *exchange property* and those with the *anti-exchange property*. Exchange alignments provide an abstract framework for much of vector space theory, including the concepts of *hyperplane* and *dimension*. Exchange alignments on a finite set are well-known under the name of *matroids*. Anti-exchange alignments, which are of more recent vintage, in the same way provide an abstract framework for some aspects of convexity theory. We also study properties of anti-exchange alignments on a finite set, which are here called *antimatroids*.

1 PARTIALLY ORDERED SETS

Although we will make no use of the theory of partially ordered sets and lattices, we will at times use some of the concepts of these subjects and for convenience of reference we state some definitions here.

A set X is said to be *partially ordered* if a binary relation $\leq$ is defined on X with the properties

(O1) $x \leq x$ (reflexivity),
(O2) *if* $x \leq y$ *and* $y \leq x$, *then* $x = y$ (antisymmetry),
(O3) *if* $x \leq y$ *and* $y \leq z$, *then* $x \leq z$ (transitivity).

Here a *binary relation* on X is just a subset R of the product set $X \times X$ and $x \leq y$ denotes that $(x,y) \in R$. It may be that neither $x \leq y$ nor $y \leq x$, in which case x and y are said to be *incomparable*. A partially ordered set is *totally ordered* if no two elements are incomparable.

If $x \leq y$ and $x \neq y$, we write $x < y$. Instead of $x \leq y$ we may write $y \geq x$, and instead of $x < y$ we may write $y > x$.

We say that $x \in X$ is an *upper bound* for a subset Y of a partially ordered set X if $y \leq x$ for every $y \in Y$, and a *lower bound* for Y if $x \leq y$ for every $y \in Y$. An upper bound for Y is said to be a least upper bound, or *supremum*, for Y if it is a lower bound for the set of all upper bounds. Similarly a lower bound for Y is said to be a greatest lower bound, or *infimum*, for Y if it is an upper bound for the set of all lower bounds. It follows from **(O2)** that Y has at most one supremum and at most one infimum.

A partially ordered set is a *lattice* if any two elements x,y have a supremum $x \vee y$ and an infimum $x \wedge y$. It is a *complete lattice* if every subset Y has a supremum and an infimum.

If X and Y are partially ordered sets, a map $f: X \to Y$ is said to be *order-preserving* if $x_1 \leq x_2$ implies $f(x_1) \leq f(x_2)$. It is said to be an *order isomorphism* if, in addition, it is a bijection and $f(x_1) \leq f(x_2)$ implies $x_1 \leq x_2$. An order isomorphism between two lattices is a *lattice isomorphism*.

The preceding definitions are all rather basic. A few concepts of lesser importance will be defined when they are encountered.

For us the most important example of a partially ordered set is the collection X of all subsets of a given set C, where $A \leq B$ denotes that the subset A is contained in the subset B. This partially ordered set is in fact a complete lattice, since the supremum of any family $\{A_\alpha\}$ of subsets is their union $\bigcup A_\alpha$ and the infimum is their intersection $\bigcap A_\alpha$.

One property of partially ordered sets will be proved here, since it will be used repeatedly. This property may be formulated in several equivalent ways. We choose here a formulation due to Hausdorff (1927) and will always appeal to it. Other formulations, including the well-ordering theorem and the popular Zorn's lemma, may be found in Hewitt and Stromberg (1975).

HAUSDORFF'S MAXIMALITY THEOREM *Every nonempty partially ordered set X contains a maximal totally ordered subset.*

Proof Let $\mathcal{A}$ be the family of all totally ordered subsets of X. Evidently $\mathcal{A}$ is not empty, since it contains any *singleton*, i.e. any subset of X containing exactly one element. If $\mathcal{T}$ is any subfamily of $\mathcal{A}$ which is totally ordered by inclusion, then the union of all the (totally ordered) sets in $\mathcal{T}$ is again totally ordered and hence is in $\mathcal{A}$.

By the axiom of choice there exists a function f which associates to each nonempty subset E of X an element $f(E)$ of E. For any $A \in \mathcal{A}$, let $A^\dagger$ be the set of all $x \in X \backslash A$ such that $A \cup x \in \mathcal{A}$. We put

$$g(A) = \begin{cases} A \cup f(A^\dagger) & \text{if } A^\dagger \neq \emptyset, \\ A & \text{if } A^\dagger = \emptyset. \end{cases}$$

The function g: $\mathcal{A} \to \mathcal{A}$ has the property that $A \subseteq g(A)$ and that at most one element of $g(A)$ is not in A. We wish to show that $g(A) = A$ for at least one $A \in \mathcal{A}$, since then $A^\dagger = \emptyset$ and A is a maximal element of $\mathcal{A}$.

Fix $A_0 \in \mathcal{A}$. We will call a subfamily $\mathcal{B}$ of $\mathcal{A}$ a 'tower' if it has the following three properties:

(i) $A_0 \in \mathcal{B}$,
(ii) if $\mathcal{T}$ is any subfamily of $\mathcal{B}$ which is totally ordered by inclusion, then the union of all the sets in $\mathcal{T}$ is again in $\mathcal{B}$,
(iii) if $B \in \mathcal{B}$, then $g(B) \in \mathcal{B}$.

For example, the family of all $B \in \mathcal{A}$ such that $A_0 \subseteq B$ is a tower. If $\mathcal{B}_0$ is the intersection of all towers, then $\mathcal{B}_0$ is itself a tower but no proper subfamily of $\mathcal{B}_0$ is a tower. Evidently $A_0 \subseteq B$ for every $B \in \mathcal{B}_0$. It is enough to show that $\mathcal{B}_0$ is totally ordered by inclusion. For the union $\overline{B}$ of all sets in $\mathcal{B}_0$ will be in $\mathcal{B}_0$, by (ii), and $g(\overline{B}) \in \mathcal{B}_0$, by (iii). Hence $g(\overline{B}) \subseteq \overline{B}$, by the definition of $\overline{B}$. Since $\overline{B} \subseteq g(\overline{B})$, by the definition of g, it follows that $g(\overline{B}) = \overline{B}$.

Let $\mathcal{C}$ be the family of all $C \in \mathcal{B}_0$ such that, for every $B \in \mathcal{B}_0$, either $B \subseteq C$ or $C \subseteq B$. For example, $A_0 \in \mathcal{C}$. For each $C \in \mathcal{C}$, let $\mathcal{D}(C)$ be the family of all $B \in \mathcal{B}_0$ such that either $B \subseteq C$ or $g(C) \subseteq B$. Again, $A_0 \in \mathcal{D}(C)$. Thus (i) is satisfied by $\mathscr{C}$ and by each $\mathscr{D}(C)$. Evidently also (ii) is satisfied by $\mathscr{C}$ and by each $\mathscr{D}(C)$.

To show that $\mathscr{D}(C)$ is a tower it remains to show that $B \in \mathscr{D}(C)$ implies $g(B) \in \mathscr{D}(C)$, i.e. that $B \subseteq C$ and $g(C) \subseteq B$ each imply either $g(B) \subseteq C$ or $g(C) \subseteq g(B)$. If $g(C) \subseteq B$ then also $g(C) \subseteq g(B)$, since $B \subseteq g(B)$. If $B = C$, then $g(C) = g(B)$. If $B \subset C$, then C cannot be a proper subset of $g(B)$, since $g(B) \setminus B$ contains at most one element, and hence $g(B) \subseteq C$, since $C \in \mathscr{C}$ and $g(B) \in \mathscr{B}_0$.

Thus $\mathscr{D}(C)$ is a tower. Since $\mathscr{B}_0$ is minimal, it follows that $\mathscr{D}(C) = \mathscr{B}_0$ for each $C \in \mathscr{C}$. Thus if $B \in \mathscr{B}_0$ and $C \in \mathscr{C}$, then either $B \subseteq C$ or $g(C) \subseteq B$. Thus $g(C) \in \mathscr{C}$ and $\mathscr{C}$ is a tower. Since $\mathscr{B}_0$ is minimal, it follows that $\mathscr{C} = \mathscr{B}_0$. Hence, by the definition of $\mathscr{C}$, $\mathscr{B}_0$ is totally ordered. □

2 ALIGNED SPACES

Let X be a set and let $\mathscr{C}$ be a collection of subsets of X. The collection $\mathscr{C}$ will be said to be an *alignment* on X if it has the following three properties:

(A1) *the set X is itself in $\mathscr{C}$,*

(A2) *the intersection of any nonempty family of sets in $\mathscr{C}$ is again a set in $\mathscr{C}$,*

(A3) *the union of any nonempty family of sets in $\mathscr{C}$ which is totally ordered by inclusion is again a set in $\mathscr{C}$.*

For example, the collection of all subgroups of a group G is an alignment on G, and the collection of all subspaces of a vector space V is an alignment on V.

The subsets of X which are in $\mathscr{C}$ will be said to be *convex*. In subsequent chapters we will study particular types of alignment, and the notion of convex set will be correspondingly restricted.

For any set $S \subseteq X$, let $[S]$ denote the intersection of all convex sets which contain S. Then, by **(A1)–(A2)**, $[S]$ is itself a convex set, which we call the *convex hull* of S.

PROPOSITION 1 *Convex hulls have the following properties*:

(H1) $S \subseteq [S]$,

(H2) $S \subseteq T$ *implies* $[S] \subseteq [T]$,

(H3) $[[S]] = [S]$,

(H4) *the convex hull of any set is the union of the convex hulls of all its finite subsets.*

Proof Properties **(H1)** and **(H2)** follow immediately from the definition of convex hull, and **(H3)** restates that $[S]$ is convex. The proof of **(H4)** requires the use of **(A3)** and is not so immediate.

A set $S \subseteq X$ will be said to be 'good' if $[S] = \bigcup [F]$, where F runs through all finite subsets of S, and will be said to be 'excellent' if $S \cup H$ is good for every finite set $H \subseteq X$. Thus every finite subset of X is excellent, and $S \cup H$ is excellent if S is excellent and H finite.

Let S be an arbitrary nonempty subset of X and let $\mathscr{E}$ be the family of all excellent subsets of S. Then $\mathscr{E}$ is not empty, since it contains every finite subset of S. By Hausdorff's maximality theorem the family $\mathscr{E}$, partially ordered by inclusion, contains a maximal totally ordered subfamily $\mathscr{T}$. Put $M = \bigcup_{T \in \mathscr{T}} T$. We are going to show that $M \in \mathscr{E}$.

If H is any finite subset of X, the family $\{[T \cup H]: T \in \mathscr{T}\}$ is also totally ordered by inclusion. Hence if we put $C_H = \bigcup_{T \in \mathscr{T}} [T \cup H]$, then C_H is convex. Since $M \cup H \subseteq C_H$, it follows that $[M \cup H] \subseteq C_H$. On the other hand, $T \subseteq M$ implies $[T \cup H] \subseteq [M \cup H]$ and hence $C_H \subseteq [M \cup H]$. Thus $C_H = [M \cup H]$. Consequently, if $x \in [M \cup H]$ then $x \in [T \cup H]$ for some $T \in \mathscr{T}$. Since T is excellent, it follows that $x \in [F \cup H]$ for some finite set $F \subseteq T \subseteq M$. Thus $M \cup H$ is good and M is excellent, as we wished to prove.

By the definition of $\mathscr{T}$, no excellent subset of S properly contains M. Since $M \cup s$ is excellent if $s \in S \backslash M$, it follows that $M = S$. Thus S is indeed good. □

The properties **(H1)–(H3)** alone imply that, for arbitrary sets $S, T \subseteq X$,

$$[S \cup T] = [S \cup [T]] = [[S] \cup [T]].$$

Let 2^X be the collection of all subsets of the set X. A map $h: 2^X \to 2^X$ is said to be a *hull operator* on X if it has the following three properties:

(i) $S \subseteq h(S)$,

(ii) $S \subseteq T$ *implies* $h(S) \subseteq h(T)$,

(iii) $h(h(S)) = h(S)$.

The hull operator is said to be *algebraic* if, in addition,

(iv) *$x \in h(S)$ implies $x \in h(F)$ for some finite set $F \subseteq S$.*

We have shown that from any alignment on X we can derive an algebraic hull operator. However, the process can also be reversed. If $h: 2^X \to 2^X$ is an algebraic hull operator on X, then one sees immediately that the collection $\mathscr{C}$ of all sets $S \subseteq X$ such that $h(S) = S$ is an alignment on X. In fact, there is a bijective correspondence between alignments on X and algebraic hull operators on X. For our purposes it is more convenient to take alignments as the starting-point.

An alignment $\mathscr{C}$ on a set X may be said to be *normed* if it has the additional property

(A0) *the empty set $\emptyset$ is in $\mathscr{C}$.*

There is no loss of generality in restricting attention to normed alignments. For let $\mathscr{C}$ be an arbitrary alignment on X, and put $E = \bigcap_{C \in \mathscr{C}} C$. Then the collection $\mathscr{D}$ of all sets $\{C \setminus E: C \in \mathscr{C}\}$ is a normed alignment on X, since $E \in \mathscr{C}$.

We will assume throughout the remainder of this section that we are given a set X and a normed alignment $\mathscr{C}$ on X. We will say that the pair $(X,\mathscr{C})$ is an *aligned space*, or simply that X is an aligned space if the meaning is clear. The qualification 'nonempty' in the statement of **(A2)** and **(A3)** may now be omitted, and convex hulls now have not only the properties **(H1)**–**(H4)** of Proposition 1 but also the property

(H0) $[\emptyset] = \emptyset$.

The collection $\mathscr{C}$ of all convex sets, partially ordered by inclusion, is a *complete lattice.* For any family $\{C_\alpha\}$ of convex sets has an infimum, namely $\bigcap C_\alpha$, and a supremum, namely the intersection of all convex sets which contain $\bigcup C_\alpha$.

If $S \subseteq X$ and $e \in S$, then e is said to be an *extreme point* of the set S if $e \notin [S \setminus e]$. Clearly if e is an extreme point of S, then it is also an extreme point of every subset of S which contains it. From the definition we obtain also

PROPOSITION 2 *Let C be a convex set and $e \in C$. Then e is an extreme point of C if and only if $C \setminus e$ is convex.*

Proof If $C \setminus e$ is convex then $[C \setminus e] = C \setminus e$, and if $C \setminus e$ is not convex then $[C \setminus e] = C$. □

Since any intersection of convex sets is again a convex set, it follows from Proposition 2 that if an arbitrary collection of extreme points is removed from a convex set, then the remaining set is convex.

We will denote by $\mathbf{E}(S)$ the set of all extreme points of the set S. It can be characterized in the following way:

PROPOSITION 3 *If $S \subseteq X$, then $\mathbf{E}(S)$ is the intersection of all subsets of S which have the same convex hull as S.*

Proof If e is an extreme point of S and if T is a subset of S with $[T] = [S]$, then $e \in T$, since $[S \setminus e]$ is a proper subset of $[S]$. On the other hand, if $e \in S$ is not an extreme point of S then $e \in [S \setminus e]$. Hence $S \subseteq [S \setminus e]$ and $[S] = [S \setminus e]$. But $e \notin S \setminus e$. □

A set $S \subseteq X$ will be said to be *independent* if every point of S is an extreme point of S.

It follows at once from this definition that, for an arbitrary set $S \subseteq X$, the set $\mathbf{E}(S)$ of all extreme points of S is an independent set. Furthermore, any singleton is an independent set and any subset of an independent set is again an independent set. On the other hand, it follows from **(H4)** that an infinite set is independent if every finite subset is independent. Consequently the union of a totally ordered collection of independent sets is again an independent set. Hence, by Hausdorff's maximality theorem, any independent subset of a set is contained in a maximal independent subset.

A subset T of a set S will be said to *generate* S if $[T] = [S]$, and will be said to be a *basis* of S if in addition it is independent.

It follows at once from the definitions that a subset T of a set S is a basis of S if and only if T generates S, but no proper subset of T generates S. Hence a set has a finite basis if it is generated by a finite subset. Any basis of a set S is a maximal independent subset of S, but in general the converse is not true.

A subset A of a convex set C will be said to be a *face* of C if A is convex and if, for every set $S \subseteq C$,

$$[S] \cap A \subseteq [S \cap A].$$

It is actually sufficient to require that $[F] \cap A \subseteq [F \cap A]$ for every finite set $F \subseteq C$. For then, if $S \subseteq C$ and $x \in [S] \cap A$, there is a finite set $F \subseteq S$ such that

$$x \in [F] \cap A \subseteq [F \cap A] \subseteq [S \cap A].$$

Obviously C is itself a face. The *proper* faces of a convex set C are the faces other than C itself.

PROPOSITION 4 *The collection of all faces of a convex set C is a normed alignment on C.*

Proof Evidently $\emptyset$ and C are faces of C. If $\{A_i: i \in I\}$ is a family of faces of C and $A = \bigcap_{i \in I} A_i$, then A is a convex subset of C. Suppose $S \subseteq C$ and $x \in [S] \cap A$. Then $x \in [F]$ for some finite $F \subseteq S$. Moreover we may assume that $x \notin [F^*]$ for every proper subset F^* of F. Then, since $x \in [F] \cap A_i \subseteq [F \cap A_i]$, we have $F \cap A_i = F$ for all $i \in I$. Hence $F \subseteq A_i$ for all $i \in I$, and so $F \subseteq A$. Consequently $x \in [F] \subseteq [S \cap A]$, proving that A is a face of C.

If $\{A_i: i \in I\}$ is a family of faces of C which is totally ordered by inclusion and $A = \bigcup_{i \in I} A_i$, then A is a convex subset of C. Suppose $S \subseteq C$ and $x \in [S] \cap A$. Then, for some $i \in I$, $x \in [S] \cap A_i \subseteq [S \cap A_i]$. Hence $[S] \cap A \subseteq [S \cap A]$, and A is a face of C. □

The collection of all faces of a convex set C, partially ordered by inclusion, is also a complete lattice, since any family $\{F_\alpha\}$ of faces has both an infimum, namely $\bigcap F_\alpha$, and a supremum, namely the intersection of all faces which contain $\bigcup F_\alpha$.

PROPOSITION 5 *If A is a face of the convex set C, then $C \setminus A$ is convex.*

Proof Taking $S = C \setminus A$ in the definition of a face, we obtain $[C \setminus A] \cap A = \emptyset$. Since $[C \setminus A] \subseteq C$, it follows that $[C \setminus A] = C \setminus A$. □

PROPOSITION 6 *If A,B,C are convex sets such that A is a face of B and B is a face of C, then A is a face of C.*

Proof Suppose $S \subseteq C$ and $x \in [S] \cap A$. Then $x \in [S] \cap B$ and hence $x \in [S \cap B]$, since B is a face of C. From $S \cap B \subseteq B$ and $x \in [S \cap B] \cap A$ it follows that $x \in [S \cap A]$, since A is a face of B. □

PROPOSITION 7 *If A is a face of the convex set C and B a face of the convex set D, then $A \cap B$ is a face of the convex set $C \cap D$.*

In particular, if A is a face of the convex set C, then A is a face of any convex set B such that $A \subseteq B \subseteq C$.

Proof Certainly $E: = A \cap B$ is convex. Suppose $S \subseteq C \cap D$ and $x \in [S] \cap E$. From $S \subseteq C$ and $x \in [S] \cap A$ we obtain $x \in [S \cap A]$. From $S \cap A \subseteq D$ and $x \in [S \cap A] \cap B$ we obtain $x \in [S \cap E]$. □

PROPOSITION 8 *A singleton $\{e\}$ is a face of a convex set C if and only if $[e] = \{e\}$ and e is an extreme point of C.*

Proof If $\{e\}$ is a face of C, then $\{e\}$ is convex and, taking $S = C \setminus e$ in the definition of a face, $e \notin [C \setminus e]$.

Suppose, on the other hand, that $[e] = \{e\}$ and e is an extreme point of C. Let S be any subset of C. If $e \notin S$ then $e \notin [S]$, since $C \setminus e$ is convex. If $e \in S$, then $e \in [S]$. In either case, $[S] \cap \{e\} \subseteq [S \cap e]$. □

We will say that a set $S \subseteq X$ is a *Radon set* if for every partition

$$S = S_1 \cup S_2, \quad S_1 \cap S_2 = \emptyset,$$

we also have $[S_1] \cap [S_2] = \emptyset$. This implies that for any pair of disjoint subsets S',S'' of S we have $[S'] \cap [S''] = \emptyset$, since we can extend the pair to a partition $S_1 = S'$, $S_2 = S \setminus S'$. Consequently any subset of a Radon set is again a Radon set. On the other hand, it follows from **(H4)** that an infinite set is a Radon set if every finite subset is a Radon set. Hence the union of a totally ordered collection of Radon sets is again a Radon set, and any Radon set is contained in a maximal Radon set. Moreover any singleton is a Radon set, by **(H0)**.

We will say also that a set $S \subseteq X$ is a *Helly set* if

$$\bigcap_{s \in S} [S \setminus s] = \emptyset.$$

Any subset of a Helly set is again a Helly set. For if S is a Helly set and $T \subseteq S$, then

$$\bigcap_{t \in T} [T \setminus t] \subseteq \bigcap_{t \in T} [S \setminus t] \cap \bigcap_{s \in S \setminus T} [S \setminus s]$$
$$= \bigcap_{s \in S} [S \setminus s] = \emptyset.$$

We now show that an infinite set S is a Helly set if every finite subset is a Helly set. Indeed if S is not a Helly set, there exists a point $s_0 \in [S \setminus s]$ for every $s \in S$. Taking $s = s_0$, it follows that there exists a finite set $F_0 \subseteq S \setminus s_0$ such that $s_0 \in [F_0]$. If $F_0 = \{s_1,\ldots,s_n\}$ then, in the same way, for each $i = 1,\ldots,n$ there exists a finite set $F_i \subseteq S \setminus s_i$ such that $s_0 \in [F_i]$. It follows that the finite subset

$F = F_0 \cup F_1 \cup ... \cup F_n$ is not a Helly set. For if $s \notin F_0$ then $F_0 \subseteq S \setminus s$ and $s_0 \in [F_0] \subseteq [S \setminus s]$, whereas if $s \in F_0$ then $s = s_i$ for some $i \in \{1,...,n\}$ and hence $s_0 \in [F_i] \subseteq [S \setminus s]$.

It now follows, as for Radon sets, that the union of a totally ordered collection of Helly sets is again a Helly set, and any Helly set is contained in a maximal Helly set. Moreover, any singleton is a Helly set.

PROPOSITION 9 *Every Helly set is a Radon set, and every Radon set is an independent set.*

Proof Suppose that the set $S \subseteq X$ is not a Radon set. Then there exists a partition $S = S_1 \cup S_2$, $S_1 \cap S_2 = \emptyset$ with $[S_1] \cap [S_2] \neq \emptyset$. Let $x \in [S_1] \cap [S_2]$. If $s \in S_1$, then $S_2 \subseteq S \setminus s$ and hence $x \in [S \setminus s]$. If $s \in S_2$, then $S_1 \subseteq S \setminus s$ and again $x \in [S \setminus s]$. Thus $x \in \bigcap_{s \in S} [S \setminus s]$ and S is not a Helly set.

On the other hand, by the definition of an extreme point, every point of a Radon set R is an extreme point of R. Thus any Radon set is an independent set. □

Let S be a Helly set with $|S| = h + 1$. (Here, and later, we denote the cardinality of a set S by $|S|$.) If $s_1,...,s_{h+1}$ are the elements of S, then the family of convex sets $C_i = [S \setminus s_i]$ $(i = 1,...,h+1)$ has empty intersection, although every subfamily of h of the sets has nonempty intersection. This statement admits a converse:

PROPOSITION 10 *Suppose that some finite family $\mathscr{C}$ of at least $h + 1$ convex sets has empty intersection, although every subfamily of h of the sets has nonempty intersection. Then there exists a Helly set $S \subseteq X$ with $|S| = h + 1$.*

Proof If $\mathscr{F} \subseteq \mathscr{C}$ is a minimal subfamily with empty intersection, then $|\mathscr{F}| \geq h + 1$. For each $F \in \mathscr{F}$ there exists a point a_F which belongs to every convex set in $\mathscr{F}$ except F. Moreover, if $F' \in \mathscr{F}$ and $F' \neq F$, then $a_{F'} \neq a_F$.

If we put $A = \bigcup_{F \in \mathscr{F}} a_F$ then for every $F \in \mathscr{F}$ we have $A \setminus a_F \subseteq F$. Hence

$$\bigcap_{F \in \mathscr{F}} [A \setminus a_F] \subseteq \bigcap_{F \in \mathscr{F}} F = \emptyset.$$

Thus A is a Helly set containing at least h+1 points, and we can take S to be any subset of A containing exactly h+1 points. □

We will say that a set $S \subseteq X$ is a *Carathéodory set* if

$$[S] \neq \bigcup_{s \in S} [S \setminus s].$$

A Carathéodory set is nonempty and also finite, since the convex hull of any set is the union of the convex hulls of all its finite subsets. Moreover any Carathéodory set is an independent set, since if $s \in [S \setminus s]$ then $[S] = [S \setminus s]$. Evidently also any singleton is a Carathéodory set. However, a nonempty subset of a Carathéodory set need not again be a Carathéodory set; an example is given at the end of Chapter II.

For any set $S \subseteq X$ and any $x \in [S]$, there exists a Carathéodory set $F \subseteq S$ such that $x \in [F]$. For there exists a finite set $F \subseteq S$ such that $x \in [F]$, and we can choose F so that no proper subset has the same property.

Finally we consider some ways of manufacturing new aligned spaces from given ones.

(i) *Restriction*: Let $\mathscr{C}$ be a normed alignment on the set X and let Y be an arbitrary subset of X. Then the collection $\mathscr{D}$ of all sets $C \cap Y$, where $C \in \mathscr{C}$, is a normed alignment on Y.

If $S \subseteq Y$, then $[S]_{\mathscr{D}} = [S]_{\mathscr{C}} \cap Y$. Hence the extreme points of S in the aligned space $(Y,\mathscr{D})$ are the same as the extreme points of S in the aligned space $(X,\mathscr{C})$. It follows that a subset of Y is independent in $(Y,\mathscr{D})$ if and only if it is independent in $(X,\mathscr{C})$.

(ii) *Join*: Let X_1, X_2 be disjoint sets and let $\mathscr{C}_k$ be a normed alignment on X_k $(k = 1,2)$. If $X = X_1 \cup X_2$ and $\mathscr{C} = \{C: C = C_1 \cup C_2, \text{ where } C_k \in \mathscr{C}_k \ (k = 1,2)\}$, then $\mathscr{C}$ is a normed alignment on X.

If $S \subseteq X$, then the extreme points of S in the aligned space $(X,\mathscr{C})$ are the extreme points of $S \cap X_1$ in the aligned space $(X_1,\mathscr{C}_1)$, together with the extreme points of $S \cap X_2$ in the aligned space $(X_2,\mathscr{C}_2)$. Hence S is independent in $(X,\mathscr{C})$ if and only if $S \cap X_k$ is independent in $(X_k,\mathscr{C}_k)$ for $k = 1,2$.

(iii) *Contraction*: Let $\mathscr{C}$ be an alignment on the set X and let Y be an arbitrary subset of X. Then the collection $\mathscr{D}$ of all subsets D of Y such that

$$[D \cup (X \setminus Y)] \cap Y = D$$

is an alignment on Y. The alignment $\mathscr{D}$ is normed if and only if $X \setminus Y \in \mathscr{C}$.

(iv) *Intersection*: Let $\mathscr{C}$ and $\mathscr{D}$ be alignments on the same set X. It is easily verified that $h(S) = [S]_{\mathscr{C}} \cap [S]_{\mathscr{D}}$ is an algebraic hull operator on X. It follows that the collection $\mathscr{E}$ of all sets $C \cap D$, where $C \in \mathscr{C}$ and $D \in \mathscr{D}$, is again an alignment on X, which is normed if either $\mathscr{C}$ or $\mathscr{D}$ is normed.

3 ANTI-EXCHANGE ALIGNMENTS

An aligned space X will be said to have the *anti-exchange property*, and its alignment will be called an *anti-exchange alignment*, if the following additional axiom is satisfied:

(Æ) *for any distinct $x,y \in X$ and any $S \subseteq X$, if $y \in [S \cup x]$ but $y \notin [S]$, then $x \notin [S \cup y]$.*

This definition can be reformulated in several other ways:

PROPOSITION 11 *An aligned space X has the anti-exchange property if and only if one of the following equivalent properties holds:*

(i) *a subset C of a convex set D is a maximal proper convex subset of D if and only if $D \setminus C = \{e\}$, where e is an extreme point of D;*
(ii) *any set $S \subseteq X$ has the same extreme points as its convex hull* $[S]$;
(iii) *for any convex set $C \subset X$ and any point $x \in X \setminus C$, the set $[C \cup x] \setminus x$ is convex;*
(iv) *for any finite set $S \subseteq X$,* $[S] = [\mathrm{E}(S)]$.

Proof Let C be a subset of the convex set D. If $|D \setminus C| = 1$ then, by Proposition 2, C is a maximal proper convex subset of D if and only if $D \setminus C = \{e\}$, where e is an extreme point of D. Suppose $|D \setminus C| > 1$ and let x,y be distinct points of $D \setminus C$. If $y \notin [C \cup x]$ then $E := [C \cup x]$ is a convex set such that $C \subset E \subset D$, and if $x \notin [C \cup y]$ then $E' := [C \cup y]$ is a convex set such that $C \subset E' \subset D$. It follows that (Æ) $\Rightarrow$ (i).

If $S \subseteq X$ and e is an extreme point of $[S]$ then $e \in S$, by Proposition 3, and hence e is an extreme point of S. Assume (i) holds, but (ii) does not. Then, for some $S \subseteq X$ and some $e \in S$, $e \notin C := [S \setminus e]$ and $T := [S] \setminus e$ is not convex. Hence $C \subset T$ and $[S] = [T] = T \cup e$. Consequently C is not a maximal proper convex subset of $[S]$. Let $\mathfrak{D}$ be the family of all convex sets D such that $C \subset D \subset [S]$. Then $\mathfrak{D}$, partially ordered by inclusion, contains a maximal totally ordered subfamily $\mathcal{T}$. Evidently $M = \bigcup_{D\in\mathcal{T}} D$ is convex and $M \subset T$, since $e \notin M$. Consequently M is not a maximal proper convex subset of $[S]$, which contradicts the definition of $\mathcal{T}$. Thus (i) $\Rightarrow$ (ii).

If $C \subset X$ is convex and $x \in X\backslash C$, then x is an extreme point of the set $C \cup x$. If (ii) holds, then x is also an extreme point of the set $[C \cup x]$ and hence $[C \cup x]\backslash x$ is convex, by Proposition 2. Thus (ii) $\Rightarrow$ (iii).

Suppose S is a finite set such that $[\mathbf{E}(S)] \subset [S]$. Let T be a maximal subset of S containing $\mathbf{E}(S)$ such that $[T] \subset [S]$. Then there exists $x \in S\backslash[T]$. If (iii) holds, then $C = [T \cup x]\backslash x$ is convex. Since T is maximal, it follows that $[T \cup x] = [S]$ and $S \cap C = T$. Hence $S = T \cup x$ and $x \notin [S\backslash x]$. Since x is not an extreme point of S, this is a contradiction. Thus (iii) $\Rightarrow$ (iv).

Suppose that for some $x,y \in X$ and some $S \subseteq X$, $y \in [S \cup x]$ and $x \in [S \cup y]$, but $y \notin [S]$. Then there exists a finite set $F \subseteq S$ such that $y \in [F \cup x]$ and $x \in [F \cup y]$. Put $T = F \cup \{x,y\}$. If (iv) holds, then $[T] = [\mathbf{E}(T)]$. Since x and y are not extreme points of T, it follows that $[T] = [F]$. But this is a contradiction, since $[F] \subseteq [S]$ and $y \notin [S]$. Thus (iv) $\Rightarrow$ (Æ). □

Throughout the remainder of this section we assume that the aligned space X has the anti-exchange property and we derive some further simple properties.

PROPOSITION 12 *If $S \subseteq X$, then S has a basis if and only if* $[S] = [\mathbf{E}(S)]$. *Moreover* $\mathbf{E}(S)$ *is then the unique basis of S.*

Proof If T is a basis of S, then $[T] = [S]$ and every point of T is an extreme point of T. But, by Proposition 11(ii), T and S have the same extreme points. Hence $T = \mathbf{E}(S)$ and $[S] = [\mathbf{E}(S)]$.

On the other hand if $[S] = [\mathbf{E}(S)]$ then $\mathbf{E}(S)$ is a basis of S, since it is independent. □

It follows from Proposition 12 that a set has a basis if and only if its convex hull has a basis, and the bases are then the same. Also, if S has a basis and $\mathbf{E}(S) \subseteq T$, then $[S] \subseteq [T]$.

It will now be shown that if S is an independent set and $T \subseteq S$, then

$$T = [T] \cap S.$$

Indeed if $x \in [T] \cap S$, then x is an extreme point of S and hence also of $[S]$. Consequently x is an extreme point of the subset $[T]$ and hence also of T. Thus $[T] \cap S \subseteq T$, which is all that requires proof.

Conversely, a set S is independent if for each $T \subseteq S$ there exists a convex set C such that $T = C \cap S$. For suppose some $x \in S$ is not an extreme point of S. Then $x \in [S \setminus x]$. By hypothesis there exists a convex set C such that $S \setminus x = C \cap S$. Then $[S \setminus x] \subseteq C$. Hence $x \in C \cap S$, which is a contradiction.

Again, if S and T are sets such that $[S] = [T]$ and if S is finite, then $[S] = [S \cap T]$. For, by Proposition 11(ii), S and T have the same set E of extreme points. Thus $E \subseteq S \cap T$. Since $[E] = [S]$, by Proposition 11(iv), the claim follows.

It follows directly from the definition that the restriction of an aligned space X with the anti-exchange property to an arbitrary subset Y again has the anti-exchange property. Furthermore the join of two disjoint aligned spaces with the anti-exchange property again has the anti-exchange property. Finally, the intersection of two anti-exchange alignments on the same set X is again an anti-exchange alignment.

4 ANTIMATROIDS

An aligned space X with the anti-exchange property is said to be an *antimatroid* if it contains only finitely many points. Even if we were not initially interested in this case, we could be led to it by restriction of an alignment on a given set X to a finite subset Y.

We now consider some results which hold only in the finite case.

PROPOSITION 13 *Let X be a finite set and $\mathscr{C}$ a collection of subsets of X. Then $(X,\mathscr{C})$ is an aligned space if and only if*

(o) $\emptyset, X \in \mathscr{C}$,

(i) *$C,D \in \mathscr{C}$ implies $C \cap D \in \mathscr{C}$.*

Moreover $(X,\mathscr{C})$ has the anti-exchange property if and only if, in addition,

(ii) *$C \in \mathscr{C}$ and $C \neq X$ imply $C \cup x \in \mathscr{C}$ for some $x \in X \setminus C$.*

Proof Since X is finite, (o) and (i) imply that the intersection of any family of sets in $\mathscr{C}$ is again a set in $\mathscr{C}$, and it is trivially true that the union of any family of sets in $\mathscr{C}$ which is totally ordered by inclusion is again a set in $\mathscr{C}$. Hence $(X,\mathscr{C})$ is an aligned space if and only if (o) and (i) hold.

Suppose $(X,\mathscr{C})$ is an aligned space with the anti-exchange property, $C \in \mathscr{C}$ and $C \neq X$. If $C_0 := X$ then, by Proposition 11(iv), $C_0 = [\mathrm{E}(C_0)]$. Hence there exists a point $e_1 \in \mathrm{E}(C_0)\setminus C$. Then, by Proposition 2, $C_1 = C_0 \setminus e_1$ is also convex and $C \subseteq C_1$. If $C \neq C_1$, let $e_2 \in \mathrm{E}(C_1)\setminus C$. Then in the same way $C_2 = C_1 \setminus e_2$ is convex and $C \subseteq C_2$. If $C \neq C_2$ we can repeat the process. Since X is finite, there is a least positive integer k such that $C = C_k$. Then $C \cup e_k$ is convex.

Suppose now that the aligned space $(X,\mathscr{C})$ has the property (ii). If $C \in \mathscr{C}$ and $x \in X\setminus C$, then there exists a chain of sets in $\mathscr{C}$:

$$C \subset C \cup \{x_1\} \subset C \cup \{x_1,x_2\} \subset \ldots \subset C \cup \{x_1,\ldots,x_n\} = X.$$

Hence $x = x_s$ for some s with $1 \leq s \leq n$. Moreover, since $[C \cup x] \subseteq C \cup \{x_1,\ldots,x_s\}$, we have $[C \cup x] = C \cup \{x_{i_1},\ldots,x_{i_r}\}$, where $1 \leq i_l < \ldots < i_r = s$. Furthermore

$$[C \cup x]\setminus x = C \cup \{x_{i_1},\ldots,x_{i_{r-1}}\} = [C \cup x] \cap (C \cup \{x_1,\ldots,x_{s-1}\}) \in \mathscr{C}.$$

Hence the aligned space $(X,\mathscr{C})$ has the anti-exchange property, by Proposition 11(iii). □

In the statement of Proposition 13 we can evidently omit the requirement $X \in \mathscr{C}$ in (o) when (ii) holds. It is worth noting also that, although the anti-exchange property is obviously preserved under restriction of an aligned space to an arbitrary subset, it is not obvious that the property (ii) is preserved.

Antimatroids can also be characterized in the following way:

PROPOSITION 14 *Let X be a finite set and $\mathscr{C}$ a collection of subsets of X. Then $(X,\mathscr{C})$ is an aligned space with the anti-exchange property if and only if*

(o) $\varnothing, X \in \mathscr{C}$,

(i) *if $C,D \in \mathscr{C}$ and $C \not\subseteq D$, then $C\setminus c \in \mathscr{C}$ for some $c \in C\setminus D$.*

Proof Suppose first that the aligned space $(X,\mathscr{C})$ has the anti-exchange property, and let C,D be convex sets with $C \not\subseteq D$. Since C is the convex hull of its extreme points, by Proposition 11(iv), some extreme point c of C is not contained in D. Then $C \setminus c$ is convex.

Suppose next that $(X,\mathscr{C})$ has the properties (o),(i). We show first that if $C, D \in \mathscr{C}$, then $C \cap D \in \mathscr{C}$. Evidently we may assume that $C \not\subseteq D$. Then $C_1 = C \setminus c \in \mathscr{C}$ for some $c \in C \setminus D$ and $C_1 \cap D = C \cap D$. If $C_1 \subseteq D$, then $C \cap D = C_1 \in \mathscr{C}$. If $C_1 \not\subseteq D$, we can repeat the process. Since X is finite, the process must eventually terminate, yielding $C \cap D = C_j \in \mathscr{C}$.

By Proposition 13 it only remains to show that if $C \in \mathscr{C}$ and $C \neq X$, then $C \cup x \in \mathscr{C}$ for some $x \in X \setminus C$. By (i) there exists $x_1 \in X \setminus C$ such that $X_1 = X \setminus x_1 \in \mathscr{C}$. Evidently $C \subseteq X_1$. If $C = X_1$, there is nothing more to do. If $C \neq X_1$, the argument can be repeated. The process must eventually terminate, yielding $C = X_k$ and $X_{k-1} = C \cup x_k \in \mathscr{C}$. □

There is an interesting connection between this result and linguistics. Our *language* has as its *alphabet* the elements of the finite set X. The *admissible words* of our language are the finite sequences $\alpha = x_1 \ldots x_m$, where $x_i \in \mathbf{E}(X \setminus \{x_1, \ldots, x_{i-1}\})$ for $1 \leq i \leq m$. This language obviously has the properties

(o) $\varnothing$ is an admissible word;

(i) if $\alpha = x_1 \ldots x_m$ is an admissible word, then $x_i \neq x_j$ for $1 \leq i < j \leq m$;

(ii) there exists an admissible word $\alpha = x_1 \ldots x_n$ such that, for every $x \in X$, $x = x_i$ for some i.

Another property of our language is

(iii) if $\alpha = x_1 \ldots x_m$ and $\beta = y_1 \ldots y_n$ are admissible words, and if $x_i \notin \{y_1, \ldots, y_n\}$ for some i, then $\gamma = y_1 \ldots y_n x_j$ is an admissible word for some $x_j \in \{x_1, \ldots, x_m\}$.

To see this, take j to be the least i for which $x_i \notin \{y_1, \ldots, y_n\}$. Then there exist distinct $y_{i_1}, \ldots, y_{i_{j-1}} \in \{y_1, \ldots, y_n\}$ such that $x_1 = y_{i_1}, \ldots, x_{j-1} = y_{i_{j-1}}$. Since

$$X \setminus \{y_1, \ldots, y_n\} \subseteq X \setminus \{y_{i_1}, \ldots, y_{i_{j-1}}\},$$

we have $x_j \in \mathbf{E}(X \setminus \{y_1, \ldots, y_n\})$. Hence $\gamma = y_1 \ldots y_n x_j$ is an admissible word.

A language with the properties (o)–(iii) is sometimes called a *shelling structure*. It follows from Proposition 14 that, conversely, if X is the finite alphabet of a shelling structure then an alignment with the anti-exchange property is obtained by taking $\mathscr{C}$ to be the collection of all sets $C = X \setminus \{x_1,...,x_m\}$, where $\alpha = x_1 \dots x_m$ is an admissible word of the shelling structure.

Another result for finite aligned spaces with the anti-exchange property is the following:

PROPOSITION 15 *If X is an antimatroid and if $S \subseteq X$ is a Helly set, then there exists a Helly set C with $|C| = |S|$ which is also convex.*

Proof Obviously we may assume that S itself is not convex. For each $x \in [S] \setminus S$ we have $x \notin [S \setminus s]$ for some $s \in S$. Choose such an x so that the number of $s \in S$ for which $x \notin [S \setminus s]$ is a minimum and let e be one such s. Then e is an extreme point of S, since $x \in [S]$.

Let $T = x \cup (S \setminus e)$, so that $T \subseteq X$ and $|T| = |S|$. Since $[S] \setminus e$ is convex, we have $[T] \subseteq [S] \setminus e$. We claim that T is a Helly set. To establish this we will show that for each $y \in [T]$ we have $y \notin [T \setminus t]$ for some $t \in T$. If $y \notin [S \setminus e]$ we can take $t = x$, since $T \setminus x = S \setminus e$.

Thus we now suppose that $y \in [S \setminus e]$. Assume first that for some $t \in S$ we have $y \notin [S \setminus t]$. If $x \in [S \setminus t]$, then $t \in T$ and x is the only element of $T \setminus t$ which is not in $S \setminus t$. Since $x \in [S \setminus t]$, it follows that $[T \setminus t] \subseteq [S \setminus t]$ and hence $y \notin [T \setminus t]$.

Hence we may now assume that $x \notin [S \setminus t]$ for every $t \in S$ for which $y \notin [S \setminus t]$. Since $x \notin [S \setminus e]$ and $y \in [S \setminus e]$ this implies, by the choice of x, that $y \in S$.

Since $y \in S$ and S is a Helly set, $y \notin [S \setminus t]$ for some $t \in S$. Evidently this implies that $y = t$ and that y is an extreme point of S. Moreover $y \neq e$, since $y \in [S \setminus e]$, and hence $y \in T$. If we put $D = [S \setminus y]$, then $x \notin D$ but $x \in [S] = [y \cup D]$. Hence, by the anti-exchange property, $y \notin [x \cup D]$ and, *a fortiori*, $y \notin [T \setminus y]$. Thus our claim is now established.

In this way we have replaced the Helly set S by another Helly set T with the same cardinality but with $[T] \subset [S]$. If T is not convex we can repeat the argument. Since X is finite, we must eventually arrive at a convex Helly set which has the same cardinality as S. □

It should be noted that, in any aligned space, a convex set is a Helly set if and only if it is independent, by the definitions and Proposition 9.

5 EXCHANGE ALIGNMENTS

Throughout this section the convex sets of the aligned space X will be called *affine sets*. Furthermore the convex hull of a set $S \subseteq X$ will be called its *affine hull* and will be denoted by $\langle S \rangle$, rather than by $[S]$. The reason for these changes in terminology and notation is that in Chapter III, where the results of this section will first be used, we will be concerned with a set X which is equipped with two different alignments and it will be necessary to distinguish between them.

An aligned space X will be said to have the *exchange property*, and its alignment will be said to be an *exchange alignment*, if the following additional axiom is satisfied:

(E) *for any $x,y \in X$ and any $S \subseteq X$, if $y \in \langle S \cup x \rangle$ but $y \notin \langle S \rangle$, then $x \in \langle S \cup y \rangle$.*

This definition can be reformulated in several other ways:

PROPOSITION 16 *An aligned space X has the exchange property if and only if one of the following equivalent properties holds:*

(i) *for any $S \subset X$ and any $x \in X \setminus \langle S \rangle$, if S is an independent set, then $S \cup x$ is also an independent set;*
(ii) *for any $S \subseteq X$, if T is a maximal independent subset of S, then T is a basis of S;*
(iii) *for any $S \subseteq X$, if T is an independent subset of S, then S has a basis B such that $T \subseteq B$;*
(iv) *for any affine set $A \subset X$ and any $x \in X \setminus A$, there is no affine set A' such that $A \subset A' \subset \langle A \cup x \rangle$.*

Proof Suppose first that S is an independent set and $x \notin \langle S \rangle$. If $S \cup x$ is not independent then, for some $y \in S \cup x$, $y \in \langle (S \cup x) \setminus y \rangle$. In fact $y \in S$, since $x \notin \langle S \rangle$, and $y \notin \langle S \setminus y \rangle$, since S is independent. If X has the exchange property it follows that $x \in \langle (S \setminus y) \cup y \rangle = \langle S \rangle$, which is a contradiction. This proves that (E) $\Rightarrow$ (i).

Suppose next that T is a maximal independent subset of the set S. If (i) holds, then $S \subseteq \langle T \rangle$ and hence $\langle S \rangle = \langle T \rangle$. Thus T generates S and actually, since it is independent, T is a basis of S. Thus (i) $\Rightarrow$ (ii).

Suppose now that T is an independent subset of the set S. The family $\mathcal{F}$ of all independent subsets of S which contain T, partially ordered by inclusion, contains a maximal totally ordered subfamily $\mathcal{T}$. The union B of all the sets in $\mathcal{T}$ is an independent set containing T. Moreover B is a maximal independent subset of S. Hence (ii) $\Rightarrow$ (iii).

Suppose A is an affine set, $x \notin A$, and there exists an affine set A' such that $A \subset A' \subset \langle A \cup x \rangle$. If (iii) holds, then A has a basis B and A' has a basis $B' \supset B$. Furthermore $B' \cup x$ has a basis B'' containing B'. Since $\langle B' \rangle = A'$ and $\langle B'' \rangle = \langle A \cup x \rangle$, we must have $B'' = B' \cup x$. If $y \in B' \setminus B$, then $y \notin \langle (B' \cup x) \setminus y \rangle$, since $B' \cup x$ is independent. But $B \cup x \subseteq (B' \cup x) \setminus y$ and $\langle B \cup x \rangle = \langle A \cup x \rangle$. Since $y \in \langle A \cup x \rangle$, this is a contradiction. This proves that (iii) $\Rightarrow$ (iv).

Finally suppose $y \in \langle S \cup x \rangle$ but $y \notin \langle S \rangle$. Then $x \notin \langle S \rangle$. If (iv) holds, then $\langle S \cup y \rangle = \langle S \cup x \rangle$ and hence $x \in \langle S \cup y \rangle$. Thus (iv) $\Rightarrow$ (E). □

The property (iv) is known as the *covering property*, since in any partially ordered set (in our case, the lattice of affine sets) an element B is said to *cover* an element A if $A < B$, but there is no element C such that $A < C < B$.

Throughout the remainder of this section we assume that the aligned space X has the exchange property. It follows from Proposition 16(ii) that any subset of X has a basis. Furthermore,

PROPOSITION 17 *If T and T' are subsets of S such that T is independent and T' generates S, then there is a subset T'' of T' such that $T \cup T''$ is a basis of S.*

Proof Evidently $\langle T \cup T' \rangle = \langle S \rangle$. Let $\mathcal{F}$ be the collection of all independent sets U such that $T \subseteq U \subseteq T \cup T'$. Then $\mathcal{F}$ is not empty, since it contains T. If we partially order $\mathcal{F}$ by inclusion then, by Hausdorff's maximality theorem, $\mathcal{F}$ contains a maximal totally ordered subcollection $\mathcal{V}$. The union V of all sets in $\mathcal{V}$ is an independent set such that $T \subseteq V \subseteq T \cup T'$. Thus $V = T \cup T''$, where $T'' \subseteq T'$ and $T \cap T'' = \emptyset$. Since V is a maximal independent subset of $T \cup T'$, we have $\langle V \rangle = \langle T \cup T' \rangle$, by Proposition 16(ii). □

PROPOSITION 18 *If B_1 and B_2 are bases of a set S, and if $b_1 \in B_1 \backslash B_2$, then there exists $b_2 \in B_2 \backslash B_1$ such that $(B_1 \backslash b_1) \cup b_2$ is also a basis of S.*

Proof Since B_1 is independent, $b_1 \notin \langle B_1 \backslash b_1\rangle$ and hence, since $\langle B_2\rangle = \langle S\rangle$, there exists $b_2 \in B_2$ such that $b_2 \notin \langle B_1 \backslash b_1\rangle$. Since $b_2 \in \langle(B_1 \backslash b_1) \cup b_1\rangle$, it follows from the exchange property that $b_1 \in \langle B_3\rangle$, where $B_3 = (B_1 \backslash b_1) \cup b_2$. Since also $B_1 \backslash b_1 \subseteq \langle B_3\rangle$, it follows that $\langle B_3\rangle = \langle S\rangle$.

It remains to show that B_3 is independent. Assume on the contrary that, for some $b_3 \in B_1 \backslash b_1$, we have $b_3 \in \langle(B_1 \backslash \{b_1,b_3\}) \cup b_2\rangle$. Since $b_3 \notin \langle B_1 \backslash \{b_1,b_3\}\rangle$, it follows from the exchange property again that $b_2 \in \langle B_1 \backslash b_1\rangle$, which is a contradiction. □

An affine set $H \subset X$ is said to be a *hyperplane* (of X) if X covers H, i.e. if H is a maximal proper affine subset of X. Equivalently, an affine set $H \subset X$ is a hyperplane if $\langle H \cup x\rangle = X$ for every $x \in X \backslash H$.

It follows that if H_1,H_2 are distinct hyperplanes, then $H_1 \not\subseteq H_2$. Furthermore, if B' is a basis of a hyperplane H and $x \in X \backslash H$, then $B' \cup x$ is a basis of X.

We now show that if A is an affine set and $x \in X \backslash A$, then there is a hyperplane H such that $A \subseteq H$ and $x \notin H$. Consider the collection $\mathcal{S}$ of all affine subsets of X which contain A but do not contain x. If we suppose $\mathcal{S}$ partially ordered by inclusion then, by Hausdorff's maximality theorem, there exists a maximal totally ordered subcollection $\mathcal{S}_0$. The union H of all affine sets in $\mathcal{S}_0$ is an affine set which contains A but not x. Thus if $y \notin H$, then $x \in \langle H \cup y\rangle$ and hence, by (E), $y \in \langle H \cup x\rangle$. Consequently $X = \langle H \cup x\rangle$ and X covers H.

It follows that if H_1,H_2 are hyperplanes and if $y \in X \backslash (H_1 \cup H_2)$, $x \in H_1 \backslash H_2$, then there is a hyperplane H_3 such that $(H_1 \cap H_2) \cup y \in H_3$ and $x \notin H_3$. For $\langle(H_1 \cap H_2) \cup y\rangle$ is an affine set which does not contain x.

PROPOSITION 19 *If B is an independent set and $S = \{x_1,...,x_n\}$ a finite set such that $x_1 \notin \langle B\rangle$ and $x_j \notin \langle B \cup \{x_1,...,x_{j-1}\}\rangle$ for $j = 2,...,n$, then $B \cup S$ is an independent set.*

Proof The result holds for $n = 1$, by Proposition 16(i). We suppose $n > 1$ and use induction on n. Then $B \cup S'$, where $S' = \{x_1,...,x_{n-1}\}$, is independent. Since $x_n \notin \langle B \cup S'\rangle$, it follows that also $B \cup S$ is independent. □

PROPOSITION 20 *Let B be an independent set and let S,T be finite sets. If $B \cap S = \emptyset$ and $B \cup S$ is an independent set such that $<B \cup S> \subseteq <B \cup T>$, then $|S| \leq |T|$.*

Proof Let $S = \{x_1,\ldots,x_m\}$ and $T = \{y_1,\ldots,y_n\}$. Since $x_1 \in <B \cup T>$ the set $B \cup S_1$, where $S_1 = \{x_1,y_1,\ldots,y_n\}$, is not independent. Consequently, by Proposition 19, $y_{i_1} \in <B \cup \{x_1,y_1,\ldots,y_{i_1-1}\}>$ for some $y_{i_1} \in T$. Thus if we put $T_1 = x_1 \cup (T \setminus y_{i_1})$, then $y_{i_1} \in <B \cup T_1>$ and $<B \cup T_1> = <B \cup T>$. Similarly the set $B \cup S_2$, where $S_2 = \{x_2,x_1,T \setminus y_{i_1}\}$, is not independent. Hence for some $y_{i_2} \in T \setminus y_{i_1}$ we have $y_{i_2} \in <B \cup T_2>$, where $T_2 = \{x_2,x_1, T \setminus (y_{i_1} \cup y_{i_2})\}$, and $<B \cup T_2> = <B \cup T_1> = <B \cup T>$. If $n < m$, then by proceeding in this way we would obtain $<B \cup \{x_1,\ldots,x_n\}> = <B \cup T>$ and hence $x_m \in <B \cup \{x_1,\ldots,x_n\}>$, which is a contradiction. □

PROPOSITION 21 *Let A be a proper affine subset of X and let $B \cup S$ be a basis of X, where B is a basis of A and S is finite.*

If $B' \cup S'$ is any basis of X such that B' is a basis of A, then S' is finite and $|S'| = |S|$.

Proof It follows at once from Proposition 20 that if $B \cup T$ is a basis of X, then T is finite and $|T| = |S|$. If B' is any basis for A then, by Proposition 17, X has a basis of the form $B' \cup T'$, where $T' \subseteq B \cup S$. Evidently $T' \cap B = \emptyset$ and hence $T' \subseteq S$. But now, in the same way, X has a basis of the form $B \cup T$, where $T \subseteq T'$. Hence, by the first part of the proof, $T = T' = S$. Moreover, if $B' \cup S'$ is a basis of X, then S' is finite and $|S'| = |S|$. □

If A is a proper affine subset of X, then X has a basis of the form $B \cup S$, where B is a basis of A. We will say that A has *finite codimension d* in X if S is finite and $|S| = d$, and that A has *infinite codimension* in X if S is infinite. By Proposition 21, these definitions do not depend on the choice of B and S. Evidently an affine set is a hyperplane if and only if it has codimension 1.

In particular we can take $A = \emptyset$. We will say that X has *dimension d* if all its bases are finite and have cardinality $d + 1$, and that it is *infinite-dimensional* if all its bases are infinite.

In these definitions we may replace X by an affine subset. If A_1 and A_2 are finite-dimensional affine sets with $A_1 \subset A_2$ then the difference in their dimensions, dim A_2 – dim A_1, is equal to the codimension of A_1 in A_2. This holds also for $A_1 = \varnothing$ if we assign to the empty set the dimension –1.

PROPOSITION 22 *If A_1 and A_2 are affine sets such that $A_1 \cap A_2$ has finite codimension n in A_2, then A_1 has finite codimension m in $\langle A_1 \cup A_2 \rangle$, where $m \leq n$.*

In particular, if A_1 and A_2 are finite-dimensional affine sets, then $A_1 \cap A_2$ and $\langle A_1 \cup A_2 \rangle$ are also finite-dimensional affine sets and

$$\dim (A_1 \cap A_2) + \dim \langle A_1 \cup A_2 \rangle \leq \dim A_1 + \dim A_2.$$

Proof Let B be a basis of $A_1 \cap A_2$. Then A_1 has a basis of the form $B \cup B_1$ and A_2 has a basis of the form $B \cup B_2$, where $|B_2| = n$. Since $\langle A_1 \cup A_2 \rangle = \langle A_1 \cup B_2 \rangle$, it follows that $\langle A_1 \cup A_2 \rangle$ has a basis of the form $B \cup B_1 \cup B_2'$, where $B_2' \subseteq B_2$. □

It follows directly from the definition that the restriction of an aligned space X with the exchange property to an arbitrary subset Y again has the exchange property. Furthermore the join of two disjoint aligned spaces with the exchange property again has the exchange property.

6 NOTES

The term *alignment* originates with Jamison-Waldner (1982). The proof of Proposition 1 given here is sketched in the dissertation of Jamison (1974).

There is also a bijective correspondence between hull operators on X and collections of subsets of X with only the properties **(A1)**–**(A2)** in the definition of an alignment, but the proof of this is almost trivial. Moreover, Sierksma (1984) shows that in a sense no generality is lost by restricting attention to algebraic hull operators and alignments.

An example of a set X and a collection $\mathscr{C}$ of subsets of X having the properties **(A1)** and **(A2)**, but not the property **(A3)**, is the set $X = \mathbb{R}$ with $\mathscr{C}$ the collection of all closed subsets of $\mathbb{R}$. If $C_n = [1/n, 1]$ $(n \in \mathbb{N})$, then $C_n \in \mathscr{C}$ and $C_1 \subseteq C_2 \subseteq \ldots$, but $\bigcup_{n \in \mathbb{N}} C_n = (0,1] \notin \mathscr{C}$.

For completeness it should be mentioned that property **(A3)** in the definition of an alignment may be replaced by

(A3)′ *the union of any nonempty directed family of sets in $\mathscr{C}$ is again a set in $\mathscr{C}$.*

Here a family $\mathscr{F}$ of sets is said to be *directed* if $A,B \in \mathscr{F}$ implies $A \cup B \subseteq C$ for some $C \in \mathscr{F}$. It is trivial that **(A3)′** ⇒ **(A3)**, and it follows from Proposition 1 that **(A1)–(A3)** ⇒ **(A3)′**. Actually **(A3)** ⇒ **(A3)′**; a direct proof is given in Erné (1984).

By omitting the convexity requirements in the definition of a face of a convex set there is obtained the more general concept of an *extreme subset* of a set. The properties of extreme subsets are discussed by Lassak (1986).

The concepts of Helly, Radon and Carathéodory sets were introduced in 1976 by Soltan, who called them *h*-, *r*- and *c*-independent sets; see Soltan (1984). In Chapter III they will be related to the classical theorems of Helly, Radon and Carathéodory for convex sets in Euclidean space. The concepts replace with advantage the older concepts of Helly, Radon and Carathéodory numbers. In our terminology the Helly (Carathéodory) number of an aligned space is the maximum cardinality of any Helly (Carathéodory) set, whereas the Radon number is one more than the maximum cardinality of any Radon set.

Some significant examples of aligned spaces with the anti-exchange property will be given in Chapter II. Further results on finite aligned spaces with the anti-exchange property are given in the surveys by Edelman and Jamison (1985) and Duchet (1987). We have preferred the name 'antimatroid' to their 'convex geometry', since the essential nature of convexity seems to us to be captured more by the axiom (C) of the next chapter, than by the axiom (Æ) of this chapter.

Shelling structures are studied by Korte and Lovász (1984), who attribute to Björner the recognition of their equivalence with antimatroids. The term 'alternative precedence structure' is preferred by Björner and Ziegler in White (1992), to avoid confusion with the shellability of polyhedral complexes considered, for example, by Ziegler (1995).

The exchange property (E) was first formulated, for real vector spaces, by Grassmann (1844), §20. *Matroids*, i.e. finite aligned spaces with the exchange property, have been extensively studied; see, for example, Welsh (1976), Oxley (1992) or White (1992). They provide a common framework for independence in algebra, incidence in geometry, partial transversals in families of subsets, and disjoint unions of trees in graphs. The resulting cross-fertilization has proved remarkably fruitful.

II
Convexity

In this chapter we introduce the notion of a *convex geometry*, which is defined by just two axioms involving *points* and sets of points called *segments*. A set is *convex* if it contains, together with any two points, the segment which they determine. The connection with the previous chapter is that the collection of all convex sets is an alignment.

A real vector space is a convex geometry, and so is a vector space over any ordered division ring. We give also other interesting examples of quite a different nature. A basic property of convex geometries is that the *convex hull* of the union of two *convex sets* C,D is the union of the convex hulls of all sets $\{c,d\}$, where $c \in C$ and $d \in D$. Already in this general setting it is possible to establish several properties which were first observed for convex sets in a real vector space. In a convex geometry also, extreme points and faces may be characterized more simply than in an arbitrary alignment.

Four more axioms are then introduced, each of which on its own ensures some useful additional property of a convex geometry. These new axioms are also satisfied in any vector space over an ordered division ring, which is the underlying reason for our interest in them. Some further properties of Helly, Radon and Carathéodory sets are established under the hypotheses of this chapter. Finally, examples are given which show the independence of the six axioms and the necessity of the axioms for some of the properties which have been established.

1 EXAMPLES

'Ordinary' convexity is defined in the following way:

EXAMPLE 1 Let X be a vector space over the field $\mathbb{R}$ of real numbers. For any $x,y \in X$, define the *segment* $[x,y]$ to be the set of all $z \in X$ which can be represented in the form $z = \lambda x + (1-\lambda)y$, where $\lambda \in \mathbb{R}$ and $0 \leq \lambda \leq 1$. Then a set $C \subseteq X$ is said to be *convex* if $x,y \in C$ implies $[x,y] \subseteq C$.

For example, the subsets of $\mathbb{R}^2$ in Figure 1 are convex, but those in Figure 2 are not.

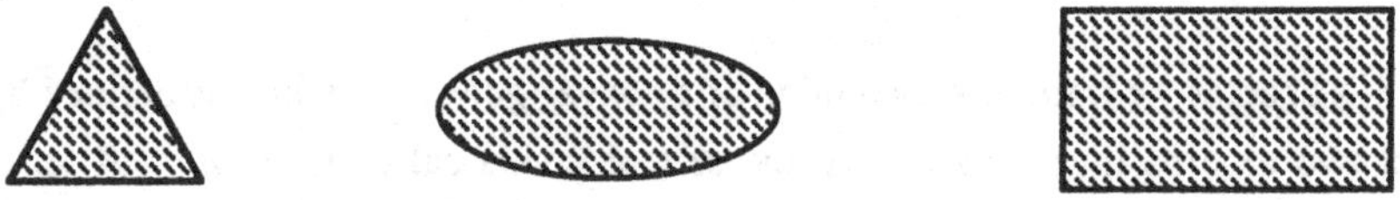

Figure 1: Some convex sets in $\mathbb{R}^2$

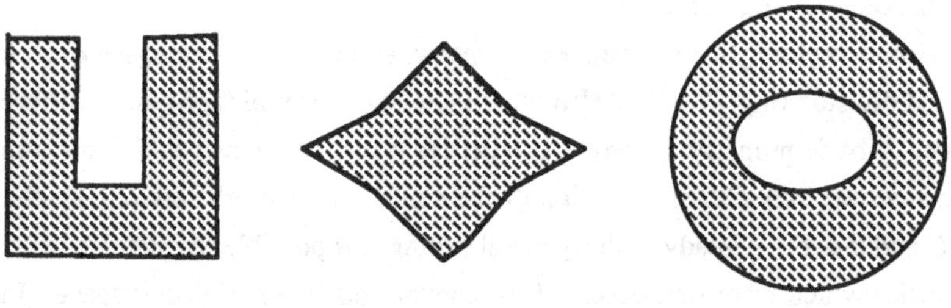

Figure 2: Some non-convex sets in $\mathbb{R}^2$

Many elementary properties of convex sets hold also in other structures, since their proofs do not make full use of the preceding definition. Here are some examples of the structures we have in mind. In each case we associate with any two elements x,y of a set X a subset $[x,y]$ of X containing them, and we then define a set $C \subseteq X$ to be *convex* if $x,y \in C$ implies $[x,y] \subseteq C$.

EXAMPLE 2 In Example 1 the field $\mathbb{R}$ of real numbers can be replaced by the field $\mathbb{Q}$ of rational numbers, leaving everything else unchanged. More generally, the field $\mathbb{R}$ can be replaced by an arbitrary ordered division ring D.

A *division ring* differs from a field only in that multiplication need not be commutative. A division ring D is *ordered* if it contains a subset P of *positive* elements which is closed under addition and multiplication and is such that D is the disjoint union of the sets $\{0\}$, P and $-P = \{-\lambda : \lambda \in P\}$. If $\lambda,\mu \in D$ we write $\lambda \leq \mu$ when either $\mu - \lambda = 0$ or $\mu - \lambda \in P$.

Any ordered field K can be embedded in a larger ordered field in the following way. Let $K(t)$ denote the field of all rational functions in one indeterminate t with coefficients from K. An element of $K(t)$ can be uniquely expressed in the form f/g, where f and g are relatively prime polynomials in t with coefficients from K and g has leading coefficient 1. If we define such an element to be 'positive' when the leading coefficient of f is positive, then $K(t)$ also acquires the structure of an ordered field. (This ordered field is 'non-archimedean', since $t > n$ for every positive integer n.)

Any ordered field K can also be embedded in an ordered division ring which is not a field. We sketch the construction, due to Hilbert, without giving detailed proofs. The set $L = L(K)$ of all formal Laurent series $a = \sum_{k \in \mathbb{Z}} \alpha_k t^k$, where $\alpha_k \in K$ and $\alpha_k \neq 0$ for at most finitely many $k < 0$, is a field if addition and multiplication are defined by

$$\sum \alpha_k t^k + \sum \beta_k t^k = \sum (\alpha_k + \beta_k) t^k, \quad \sum \alpha_k t^k \cdot \sum \beta_k t^k = \sum \gamma_k t^k,$$

where $\gamma_k = \sum_{i+j=k} \alpha_i \beta_j$. Moreover, L is an ordered field if $a = \sum_{k \in \mathbb{Z}} \alpha_k t^k$ is defined to be positive when, for some $m \in \mathbb{Z}$, $\alpha_m > 0$ and $\alpha_k = 0$ if $k < m$. If $\rho \in K$, $\rho > 0$ and $\rho \neq 1$, then the map $\psi: L \to L$ defined by

$$\psi(a) = \sum_{k \in \mathbb{Z}} \rho^k \alpha_k t^k \text{ if } a = \sum_{k \in \mathbb{Z}} \alpha_k t^k$$

is a nontrivial automorphism of the field L which preserves positivity.

Now let $\mathcal{L} = \mathcal{L}(K)$ be the set of all formal Laurent series $A = \sum_{\nu \in \mathbb{Z}} a_\nu s^\nu$, where $a_\nu \in L$ and $a_\nu \neq 0$ for at most finitely many $\nu < 0$. If we define addition and positivity as before, but define multiplication by

$$\sum a_\nu s^\nu \cdot \sum b_\nu s^\nu = \sum c_\nu s^\nu,$$

where $c_\nu = \sum_{\lambda+\mu=\nu} a_\lambda \psi^\lambda(b_\mu)$, then $\mathcal{L}$ is a noncommutative ordered division ring which contains K as a subfield.

EXAMPLE 3 In Example 1 or Example 2 we can take X to be not the whole vector space, but a given nonempty convex subset. Thus we admit only those convex sets in the vector space which are contained in X.

For example, we can take X to be the subset of $\mathbb{R}^{m^2}$ consisting of all $m \times m$ positive definite symmetric matrices.

EXAMPLE 4 Let X be a *partially ordered* set. For any $x,y \in X$, define the segment $[x,y]$ to be the set $\{x,y\}$ if x and y are incomparable, to be the set of all $z \in X$ such that $x \leq z \leq y$ if $x \leq y$, and to be the set of all $z \in X$ such that $y \leq z \leq x$ if $y \leq x$.

EXAMPLE 5 Let X be a (join) *semilattice*, i.e., a partially ordered set in which any two elements x,y have a least upper bound $x \vee y$. For any $x,y \in X$, define the segment $[x,y]$ to be the set $\{x,y,x \vee y\}$.

For example, we can take X to be the set of all subsets of a finite set E, partially ordered by inclusion. In this case the least upper bound of two subsets x,y of E is their union $x \cup y$.

EXAMPLE 6 Let X be a *tree* (i.e., a finite connected graph without circuits) and, if x,y are vertices of X, define the segment $[x,y]$ to be the set of all vertices z of X which lie on the (unique) shortest path from x to y.

EXAMPLE 7 Let X be a vector space over the field $\Lambda = \mathbb{R}(t)$ of all rational functions in one variable with real coefficients. For any $x,y \in X$ define the segment $[x,y]$ to be the set of all $z \in X$ such that $z = \lambda x + (1 - \lambda)y$, where $\lambda \in \Lambda$ and $0 \leq \lambda(t) \leq 1$ for all $t \in \mathbb{R}$ for which $\lambda(t)$ is defined (i.e., for which the denominator of λ does not vanish).

EXAMPLE 8 Let X be the set of all measurable functions $x: T \to \mathbb{R}^n$, where T is a given measure space. For any $x,y \in X$, define the segment $[x,y]$ to be the set of all $z \in X$ such that $z(t) \in \{x(t),y(t)\}$ for all $t \in T$, except a set of measure zero, i.e. $z = \chi_E x + (1 - \chi_E)y$, where χ_E is the characteristic function of some measurable set $E \subseteq T$.

In order to make it apparent that a number of properties of convex sets hold also in these structures we are going to adopt an axiomatic approach to convexity. This approach has the merit of exhibiting the logical structure of the subject and of establishing results under minimal hypotheses. It may also be argued that it forces us to formulate definitions and construct proofs in the 'right' way. After studying the general properties of 'convex geometries' we will return to the preceding examples at the end of the chapter.

2 CONVEX GEOMETRIES

Let X be a set and suppose that with any unordered pair $\{a,b\}$ of elements of X there is associated a subset $[a,b]$ of X containing a and b. The elements of X will be called *points* and the subsets $[a,b]$ *segments*. We will say that a *convex geometry* is defined on X if the following two axioms are satisfied:

(C) *if* $c \in [a,b_1]$ *and* $d \in [c,b_2]$, *then* $d \in [a,b]$ *for some* $b \in [b_1,b_2]$;
(L1) $[a,a] = \{a\}$.

Throughout this section and the next we will assume that a set X is given on which a convex geometry is defined. We are going to study the consequences of this assumption.

PROPOSITION 1 *If* $c_1,c_2 \in [a,b]$, *then* $[c_1,c_2] \subseteq [a,b]$.

Proof Let $c \in [c_1,c_2]$. From $c_1 \in [a,b]$ and $c \in [c_1,c_2]$ we obtain, by **(C)**, $c \in [a,b']$ for some $b' \in [b,c_2]$. From $c_2 \in [a,b]$ and $b' \in [b,c_2]$ we obtain, by **(C)** and **(L1)**, $b' \in [a,b]$. From $b' \in [a,b]$ and $c \in [a,b']$ we obtain, by **(C)** and **(L1)** again, $c \in [a,b]$. □

We show next that the axioms **(C)** and **(L1)** together imply a generalization of axiom **(C)**:

PROPOSITION 2 *If* $c_1 \in [a,b_1]$, $c_2 \in [a,b_2]$ *and* $c \in [c_1,c_2]$, *then* $c \in [a,b]$ *for some* $b \in [b_1,b_2]$.

Proof Applying **(C)** twice, we obtain $c \in [a,b']$ for some $b' \in [b_1,c_2]$ and $b' \in [a,b]$ for some $b \in [b_1,b_2]$. From $c \in [a,b']$ and $a,b' \in [a,b]$ we now obtain $c \in [a,b]$, by Proposition 1. □

We define a subset C of X to be *convex* if $x,y \in C$ implies $[x,y] \subseteq C$. For example, the segment $[x,y]$ is itself convex, by Proposition 1.

It follows at once from the definition that the collection $\mathscr{C}$ of all convex sets is a normed alignment on X. Thus X is an *aligned space*, and we may use the definitions and results of Chapter I, Section 2. In particular, we define the *convex hull* $[S]$ of any set $S \subseteq X$ to be the intersection of all convex sets which contain S. Our notations are consistent, since the convex hull of the set $\{x,y\}$ is the segment

$[x,y]$. However, convex geometries possess many properties not possessed by all aligned spaces, as we will now see.

PROPOSITION 3 *For any nonempty convex set C and any point* $a \notin C$,

$$[a \cup C] = \bigcup_{c \in C} [a,c].$$

Proof Suppose $x_1 \in [a,c_1]$ and $x_2 \in [a,c_2]$, where $c_1,c_2 \in C$. If $x \in [x_1,x_2]$ then, by Proposition 2, $x \in [a,c]$ for some $c \in [c_1,c_2] \subseteq C$. This proves that the set $D = \bigcup_{c \in C} [a,c]$ is convex. It is obvious that D is contained in every convex set which contains both C and a. □

Since $[x \cup [S]] = [x \cup S]$, it follows from Proposition 3 that for any nonempty set S,

$$[x \cup S] = \bigcup_{y \in [S]} [x,y].$$

In particular, *if* $x \in [a,b,c]$, *then* $x \in [a,d]$ *for some* $d \in [b,c]$. However, much more is true:

PROPOSITION 4 *For any nonempty sets* $S,T \subseteq X$,

$$[S \cup T] = \bigcup_{x \in [S], y \in [T]} [x,y].$$

Proof Put

$$R = \bigcup_{x \in [S], y \in [T]} [x,y].$$

Since $S \cup T \subseteq R \subseteq [S \cup T]$, we need only prove that R is convex. Thus we wish to show that if $z' \in [x',y']$ and $z'' \in [x'',y'']$, where $x',x'' \in [S]$ and $y',y'' \in [T]$, and if $z \in [z',z'']$, then $z \in [x,y]$ for some $x \in [S]$ and $y \in [T]$.

Since $z \in [x',y',z''] \subseteq [x',x'',y',y'']$, we have $z \in [x',w]$ for some $w \in [x'',y',y'']$. Then $w \in [x'',y]$ for some $y \in [y',y''] \subseteq [T]$. Thus $z \in [x',x'',y]$, and hence $z \in [x,y]$ for some $x \in [x',x''] \subseteq [S]$. □

In convex geometries there is a simple characterization of extreme points:

PROPOSITION 5 *If* $S \subseteq X$ *and* $e \in S$, *then e is an extreme point of S if and only if* $e \in [x,y]$, *where* $x,y \in S$, *implies* $e \in \{x,y\}$.

Proof If $e \in [x,y]$, where $x,y \in S \setminus e$, then $e \in [S \setminus e]$ and hence e is not an extreme point of S.

Suppose, on the other hand, that $e \in [x,y]$, where $x,y \in S$, implies $x = e$ or $y = e$. If e is not an extreme point of S, then $e \in [S \backslash e]$. Hence $e \in [F]$ for some finite set $F \subseteq S \backslash e$. We may assume that $e \notin [F']$ for every proper subset F' of F. If $u \in F$ then, by Proposition 4, $e \in [u,v]$ for some $v \in [F \backslash u]$. Since $v \neq e$, it follows that $u = e$. Since $u \in S \backslash e$, this is a contradiction. □

By Proposition I.8 and **(L1)**, a singleton $\{e\}$ is a face of a convex set C if and only if e is an extreme point of C. In convex geometries there is also a simple characterization of arbitrary faces. As with intervals in $\mathbb{R}$, for any $a,b \in X$ we define 'half-open' and 'open' segments by

$$[a,b) = [a,b] \backslash b, \quad (a,b] = [a,b] \backslash a, \quad (a,b) = [a,b] \backslash \{a,b\}.$$

PROPOSITION 6 *A subset A of a convex set C is a face of C if and only if A is convex and, for any $c,c' \in C$, $(c,c') \cap A \neq \emptyset$ implies $c,c' \in A$.*

Proof Let A be a convex subset of C. If $F = \{c\}$, where $c \in C$, then $[F] \cap A = [F \cap A]$ for both $c \in A$ and $c \notin A$. Suppose next that $F = \{c,c'\}$, where c,c' are distinct elements of C. Then $[F] \cap A \not\subseteq [F \cap A]$ if and only if c,c' are not both in A and $(c,c') \cap A \neq \emptyset$. Consequently, if A is a face of C then $(c,c') \cap A \neq \emptyset$ implies $c,c' \in A$.

Suppose on the other hand that $(c,c') \cap A \neq \emptyset$, where $c,c' \in C$, implies $c,c' \in A$. We wish to show that $[F] \cap A \subseteq [F \cap A]$ for any finite set $F \subseteq C$. From what has already been said, this is true if $|F| \leq 2$. Hence it is sufficient to show that if $[F] \cap A \subseteq [F \cap A]$ for some finite set $F \subseteq C$ with $|F| \geq 2$, and if $x \in C \backslash F$, then

$$[x \cup F] \cap A \subseteq [(x \cup F) \cap A].$$

By Proposition 4,

$$[x \cup F] \cap A = \bigcup_{y \in [F]} [x,y] \cap A,$$

and $[x,y] \cap A \subseteq [\{x,y\} \cap A]$. If $x \notin A$, it follows that

$$[x \cup F] \cap A \subseteq [F] \cap A \subseteq [F \cap A] = [(x \cup F) \cap A].$$

On the other hand, if $x \in A$ then

$$
\begin{aligned}
[x \cup F] \cap A &\subseteq \{x\} \cup \bigcup_{y \in [F] \cap A} [x,y] \\
&\subseteq \{x\} \cup \bigcup_{y \in [F \cap A]} [x,y] \\
&= [x \cup (F \cap A)] \\
&= [(x \cup F) \cap A]. \quad \square
\end{aligned}
$$

In an arbitrary aligned space, and even in an aligned space with the anti-exchange property, a nonempty subset of a Carathéodory set need not again be a Carathéodory set. However, convex geometries do have this desirable property:

PROPOSITION 7 *Any nonempty subset of a Carathéodory set is again a Carathéodory set.*

Proof Let S be a Carathéodory set. Since S is finite, it is enough to show that $S \setminus s$ is also a Carathéodory set for every $s \in S$. Moreover, since singletons are Carathéodory sets, we may suppose that $|S| > 2$.

Assume, on the contrary, that for some $s_0 \in S$,

$$[S \setminus s_0] = \bigcup_{s \in S \setminus s_0} [S \setminus (s \cup s_0)].$$

Then, by Proposition 4,

$$[S] = \bigcup_{t \in [S \setminus s_0]} [s_0,t] = \bigcup_{s \in S \setminus s_0} \bigcup_{t \in [S \setminus (s_0 \cup s)]} [s_0,t].$$

But $[s_0,t] \subseteq [S \setminus s]$, since $s_0 \in S \setminus s$ and $t \in [S \setminus s]$, and hence

$$[S] \subseteq \bigcup_{s \in S \setminus s_0} [S \setminus s] \subseteq \bigcup_{s \in S} [S \setminus s].$$

Since S is a Carathéodory set, this is a contradiction. $\square$

In an arbitrary convex geometry the classes of Carathéodory, Radon and Helly sets need not coincide:

EXAMPLE 9 Take $X = \mathbb{R}^3$ and for any $x = (x_1,x_2,x_3)$, $y = (y_1,y_2,y_3)$ in X define $[x,y]$ to be the set of all $z = (z_1,z_2,z_3)$ in X with $\min\{x_i,y_i\} \le z_i \le \max\{x_i,y_i\}$ $(i = 1,2,3)$. It is easily verified that with this definition X is a convex geometry. The convex subsets of X are the boxes with sides parallel to the coordinate axes. If we take $x = (0,1,2)$, $y = (1,2,0)$ and $z = (2,0,1)$, then $[x,y,z]$ is the cube consisting of all points $w = (w_1,w_2,w_3)$ with $0 \le w_i \le 2$ $(i = 1,2,3)$. The set $\{x,y,z\}$ is a Carathéodory set, since $(1/2,1/2,1/2) \notin [x,y] \cup [y,z] \cup [x,z]$, and is also a Radon set, but it is not a Helly set, since $(1,1,1) \in [x,y] \cap [y,z] \cap [x,z]$. It is easily verified also that the four points (0,0,1), (1,2,3), (3,1,2), (2,3,0) form a Radon set, but not a Carathéodory set, and that the six points $(\pm 1,0,0)$, $(0,\pm 1,0)$, $(0,0,\pm 1)$ form an independent set, but not a Radon, Helly or Carathéodory set.

For a given set $S \subseteq X$ and a given point $x \in S$, we say that a point $y \in S$ is *visible* from x if $[x,y] \subseteq S$. We denote by S_x the set of all points of S which are visible from x and we define the *kernel* of S to be the set

$$K(S) = \bigcap_{x \in S} S_x.$$

Obviously a set S is convex if and only if $K(S) = S$. A set S is said to be *star-shaped* if $K(S) \neq \emptyset$. An example in $\mathbb{R}^2$ of a set which is star-shaped, but not convex, is the middle set in Figure 2 at the beginning of the chapter.

PROPOSITION 8 *For any set $S \subseteq X$ and any $x \in S$, the kernel $K(S_x)$ of the set of all points of S visible from x is the intersection of all maximal convex subsets of S which contain x. Furthermore,*

$$K(S) = \bigcap_{x \in S} K(S_x).$$

Proof Since singletons are convex, there exists a maximal convex subset of S which contains x. Let $y \in K(S_x)$ and let M be any maximal convex subset of S which contains x. Then $M \subseteq S_x$ and hence, since $y \in K(S_x)$,

$$[y \cup M] = \bigcup_{z \in M} [y,z] \subseteq S_x.$$

Since $S_x \subseteq S$ and M is a maximal convex subset of S which contains x, it follows that $y \in M$.

Conversely, let y belong to all maximal convex subsets of S which contain x and let $z \in S_x$. Since $[x,z] \subseteq S$, there exists a maximal convex subset M of S which contains $[x,z]$. Then $y \in M$ and hence $[y,z] \subseteq M$. If $w \in [y,z]$, then $[x,w] \subseteq M$ and hence $w \in S_x$. Thus $[y,z] \subseteq S_x$. Since this holds for every $z \in S_x$, it follows that $y \in K(S_x)$.

Since $K(S_x) \subseteq S_x$, we certainly have $\bigcap_{x \in S} K(S_x) \subseteq K(S)$. It remains to show that if $y \in K(S)$, then $y \in K(S_x)$ for every $x \in S$. By the definition of kernel, $y \in S_x$ for every $x \in S$. We wish to show that $[y,z] \subseteq S_x$ for every $z \in S_x$.

If $u \in [y,z]$ then $u \in S$, since $y \in K(S)$ and $z \in S$. If $v \in [x,u]$, then $v \in [x,y,z]$ and hence $v \in [y,w]$ for some $w \in [x,z]$. Then $w \in S$, since $z \in S_x$, and hence $v \in S$, since $y \in K(S)$. Thus $[x,u] \subseteq S$ and $u \in S_x$. □

COROLLARY 9 *For any set $S \subseteq X$, the kernel $K(S)$ is the intersection of all maximal convex subsets of S. In particular, $K(S)$ is itself a convex subset of S.* □

The first statement of Corollary 9 was proved (for X a real vector space) by Toranzos (1967); the second statement was proved (for $X = \mathbb{R}^2$) by Brunn (1913).

3 ADDITIONAL AXIOMS

We now examine the consequences of imposing various additional axioms on a convex geometry. These axioms may seem arbitrary at first sight, but their natural role will become apparent in the next chapter. We first consider the axiom

(L2) *if $b \in [a,c]$, $c \in [b,d]$ and $b \neq c$, then $b \in [a,d]$.*

PROPOSITION 10 *If the convex geometry X satisfies* **(L2)**, *then the alignment of convex sets is an anti-exchange alignment.*

Proof We will show that if e is an extreme point of S, then it is also an extreme point of $[S]$. Assume on the contrary that $e \in [x,y]$, where $x,y \in [S]\setminus e$. By Proposition 4 we have $x \in [e,u]$ and $y \in [e,v]$ for some $u,v \in [S\setminus e]$. From $e \in [x,y]$, $y \in [e,v]$ and $y \neq e$ we obtain, by **(L2)**, $e \in [x,v]$. From $e \in [x,v]$, $x \in [e,u]$ and $x \neq e$ we obtain similarly $e \in [u,v]$. Since $e \notin [S\setminus e]$, this is a contradiction. □

Thus under the hypotheses of Proposition 10 all the results of Chapter I, Section 3 are valid. The axiom **(L2)** is actually essential for the convex sets of a convex geometry to form an anti-exchange alignment. Indeed if $b \in [a,c]$, $c \in [b,d]$, $b \neq c$ and $b \notin [a,d]$, then $[a,b,d]\setminus b$ contains a and c, but not b.

PROPOSITION 11 *If the convex geometry X satisfies* **(L2)** *then, for any $a,b \in X$, $(a,b]$ and (a,b) are convex sets.*

Proof We may assume $a \neq b$, by **(L1)**. Then $(a,b]$ and $[a,b)$ are convex, by Proposition 10 and Proposition I.11. Hence $(a,b) = (a,b] \cap [a,b)$ is also convex. □

Another property in which we are interested is the *no branchpoint* property:

(L3) *if* $c \notin [a,b]$ *and* $b \notin [a,c]$, *then* $[a,b] \cap [a,c] = \{a\}$.

The following proposition was proved by Hammer (1977):

PROPOSITION 12 *If the convex geometry X satisfies* **(L3)** *then, for any finite Radon set S and any subsets* S_1,S_2 *of S,*

$$[S_1] \cap [S_2] = [S_1 \cap S_2].$$

Proof The result holds by the definition of a Radon set if S_1 and S_2 are disjoint. Suppose now that $S_1 \cap S_2 = \{s\}$ is a singleton. We may clearly assume that $|S_1| > 1$, $|S_2| > 1$, and then

$$[S_1] = \cup_{x \in [S_1 \setminus s]} [s,x], \quad [S_2] = \cup_{y \in [S_2 \setminus s]} [s,y].$$

Suppose $z \in [s,x] \cap [s,y]$ for some $x \in [S_1 \setminus s]$ and $y \in [S_2 \setminus s]$. Then $x \notin [S_2]$ and $y \notin [S_1]$, because S is a Radon set. In particular, $x \notin [s,y]$ and $y \notin [s,x]$. Hence, by **(L3)**, $z = s$. Thus $[S_1] \cap [S_2] = \{s\}$ and the result holds in this case.

Suppose next that $S_1 \cap S_2$ contains $m \geq 2$ elements and assume that the result holds whenever $S_1 \cap S_2$ contains less than m elements. Write $S_1 \cap S_2 = s \cup R$, where $s \notin R$, and put $T_j = S_j \setminus (s \cup R)$ $(j = 1,2)$. Then

$$[S_1] = \cup_{x \in [R \cup T_1]} [s,x], \quad [S_2] = \cup_{y \in [R \cup T_2]} [s,y].$$

By the induction hypothesis,

$$[S_1] \cap [R \cup T_2] = [R] = [S_2] \cap [R \cup T_1].$$

Since $[R \cup T_1] \subseteq [S_1]$, it follows that

$$[R \cup T_1] \cap [R \cup T_2] = [R].$$

Hence

$$[S_1] \cap ([R \cup T_2] \setminus [R \cup T_1]) = \emptyset,$$
$$[S_2] \cap ([R \cup T_1] \setminus [R \cup T_2]) = \emptyset.$$

Suppose $z \in [s,x] \cap [s,y]$, where $x \in [R \cup T_1]$ and $y \in [R \cup T_2]$. If $x \in [R \cup T_2]$ or if $y \in [R \cup T_1]$, then $z \in [s \cup R] = [S_1 \cap S_2]$. Hence we may suppose $x \notin [R \cup T_2]$ and $y \notin [R \cup T_1]$. Then $x \notin [S_2]$ and $y \notin [S_1]$. In particular, $x \notin [s,y]$ and $y \notin [s,x]$. Hence, by **(L3)**, $z = s \in [S_1 \cap S_2]$. □

PROPOSITION 13 *If the convex geometry X satisfies* **(L3)**, *then the classes of Helly sets and Radon sets coincide.*

Proof By Proposition I.9 we need only show that every Radon set is a Helly set. Moreover we may restrict attention to finite sets, since an infinite set is a Radon (Helly) set if and only if every finite subset is a Radon (Helly) set. But if S is a finite Radon set then, by Proposition 12,

$$\bigcap_{s \in S} [S \setminus s] = [\emptyset] = \emptyset. \quad \square$$

We are also interested in the *additivity* property:

(L4) *if* $c \in [a,b]$, *then* $[a,b] = [a,c] \cup [c,b]$.

With its aid we can prove

PROPOSITION 14 *If the convex geometry X satisfies* **(L4)** *then, for any set* $S \subseteq X$ *and any* $c \in [S]$,

$$[S] = \bigcup_{s \in S} [c \cup (S \setminus s)].$$

Proof It is sufficient to prove the result for finite sets S, by **(H4)**. Suppose $S = \{s_1,\ldots,s_n\}$. The result is trivial for $n = 1$ and it holds for $n = 2$, by **(L4)**. We use induction and assume that the result holds for all finite sets containing at most n elements, where $n \geq 2$. Let $T = s_0 \cup S$ and suppose $d \in [T]$. Then, by Proposition 4, $d \in [s_0,c]$ for some $c \in [S]$. Hence, by Proposition 4 and the induction hypothesis,

$$[T] = \bigcup_{z \in [S]} [s_0,z] = \bigcup_{i=1}^{n} [c \cup (T \setminus s_i)].$$

But, by Proposition 4, $[c \cup (T \setminus s_i)]$ is the union of all segments $[x,y]$, with $x \in [s_0,c]$ and $y \in [S \setminus s_i]$. Since $[s_0,c] = [s_0,d] \cup [d,c]$, it follows from Proposition 4 again that

$$[c \cup (T \setminus s_i)] = [d \cup (T \setminus s_i)] \cup [c \cup d \cup (S \setminus s_i)] \quad (i = 1,\ldots,n).$$

Hence

$$[T] = \bigcup_{i=1}^{n} [d \cup (T \setminus s_i)] \cup \bigcup_{i=1}^{n} [c \cup d \cup (S \setminus s_i)]$$
$$= \bigcup_{i=1}^{n} [d \cup (T \setminus s_i)] \cup [d \cup S].$$

Thus the result holds also for all finite sets containing $n+1$ elements. □

PROPOSITION 15 *If the convex geometry X satisfies* **(L4)**, *then any Carathéodory set is also a Helly set.*

Proof If S is finite and there exists a point $x \in \bigcap_{s \in S} [S \setminus s]$ then, by Proposition 14,

$$[S] = \bigcup_{s \in S} [x \cup (S \setminus s)] = \bigcup_{s \in S} [S \setminus s]. \quad \square$$

PROPOSITION 16 *Suppose the convex geometry X satisfies both* **(L2)** *and* **(L4)**. *Let $S = \{s_1,...,s_n\}$ be a finite set and let d,e be distinct elements of* $[S]$. *Then* $[d,e] \subseteq [p,q]$, *where* $p,q \in \bigcup_{i=1}^{n} [S \setminus s_i]$.

Proof By Proposition 14,

$$[S] = \bigcup_{i=1}^{n} [d \cup (S \setminus s_i)] = \bigcup_{i=1}^{n} [e \cup (S \setminus s_i)].$$

Thus $e \in [d \cup (S \setminus s_j)]$ and $d \in [e \cup (S \setminus s_k)]$ for some $j,k \in \{1,...,n\}$. It follows from Proposition 4 that $e \in [d,p]$ for some $p \in [S \setminus s_j]$ and $d \in [e,q]$ for some $q \in [S \setminus s_k]$. Hence $d,e \in [p,q]$, by **(L2)**, and $[d,e] \subseteq [p,q]$, by Proposition 1. □

Finally we introduce one more axiom:

(P) *if* $c_1 \in [a,b_1]$ *and* $c_2 \in [a,b_2]$, *then* $[b_1,c_2] \cap [b_2,c_1] \neq \emptyset$.

The axiom's label is chosen in honour of Peano (1889) who, developing the work of Pasch, first used the axioms **(C)** and **(P)**. Convex geometries satisfying the axiom **(P)** possess a number of additional properties. Our first result is a counterpart to Proposition 2:

PROPOSITION 17 *Suppose the convex geometry X satisfies the axiom* **(P)**. *If* $c_1 \in [a,b_1]$, $c_2 \in [a,b_2]$ *and* $b \in [b_1,b_2]$, *then there is a point* $c \in [a,b] \cap [c_1,c_2]$.

Proof By **(P)**, there exist a point $c' \in [a,b] \cap [c_1,b_2]$ and a point $c \in [a,c'] \cap [c_1,c_2]$. Since $[a,c'] \subseteq [a,b]$, the result follows. □

PROPOSITION 18 *Suppose the convex geometry X satisfies the axiom* (**P**). *Then the following properties hold*:

(i) *if* $c \in [a,b]$ *and* $d \in [a,c]$, *then* $c \in [b,d]$;
(ii) *if* $c \in [a,b]$ *and* $b \in [a,c]$, *then* $b = c$;
(iii) *if* $c \in [a,b]$, *then* $[a,c] \cap [b,c] = \{c\}$.

Proof (i) There exists a point $e \in [b,d] \cap [c,c]$, by (**P**), and $e = c$, by (**L1**).
(ii) Take $d = b$ in (i).
(iii) If $d \in [a,c] \cap [b,c]$, then $c \in [b,d]$, by (i), and hence $d = c$, by (ii). □

The preceding two results actually hold in any aligned space satisfying the axiom (**P**). The following *sand-glass* property does require that the aligned space be derived from a convex geometry – draw a picture in $\mathbb{R}^2$ to see the reason for the name!

PROPOSITION 19 *Suppose the convex geometry X satisfies* (**P**). *If* $x \in [a,a'] \cap [b,b']$ *and* $y \in [a,b]$, *then* $x \in [y,y']$ *for some* $y' \in [a',b']$.

Proof Since $x \in [a,a']$ and $y \in [a,b]$ there exists a point $z \in [a',y] \cap [b,x]$, by (**P**). From $x \in [b,b']$ and $z \in [b,x]$ we obtain, by Proposition 18(i), $x \in [b',z] \subseteq [y,a',b']$. Hence $x \in [y,y']$ for some $y' \in [a',b']$. □

The sand-glass property in turn implies the following general statement:

PROPOSITION 20 *Suppose the convex geometry X satisfies* (**P**). *If* $c \in [a,b]$ *then, for any set* $S \subseteq X$,

$$[a \cup c \cup S] \cap [b \cup c \cup S] = [c \cup S].$$

Proof Obviously the right side is contained in the left. Suppose, on the other hand, that x is any element of the left side. Then $x \in [d,y]$ for some $d \in [a,c]$ and $y \in [S]$, and $x \in [e,z]$ for some $e \in [b,c]$ and $z \in [S]$. By Proposition 18(i), $c \in [b,d]$ and also $c \in [d,e]$. Consequently, by Proposition 19, $x \in [c,w]$ for some $w \in [y,z] \subseteq [S]$. □

As an application we prove

PROPOSITION 21 *Suppose the convex geometry X satisfies* (**P**). *If* $p \in [a_1,\ldots,a_n]$ *and* $b_i \in [p,a_i]$ $(i = 1,\ldots,n)$, *then* $p \in [b_1,\ldots,b_n]$.

Proof Assume $p \in [b_1,...,b_{k-1},a_k,...,a_n]$ for some k. Then, by Proposition 20,

$$p \in [p,b_1,...,b_k,a_{k+1},...,a_n] \cap [b_1,...,b_k,a_k,...,a_n] = [b_1,...,b_k,a_{k+1},...,a_n].$$

Since the assumption holds for $k = 1$, the result follows by induction on k. □

A basic property of convex geometries with the property **(P)** is the following *separation theorem*:

PROPOSITION 22 *Suppose the convex geometry X satisfies* **(P)**. *If C and D are disjoint convex subsets of X, then there exist disjoint convex sets C' and D' with $C' \cup D' = X$ such that $C \subseteq C'$, $D \subseteq D'$.*

Proof Let $\mathcal{F}$ be the family of all convex sets C'' which contain C but are disjoint from D. Then $\mathcal{F}$ is nonempty, since it contains C. If we partially order $\mathcal{F}$ by inclusion then, by Hausdorff's maximality theorem, $\mathcal{F}$ contains a maximal totally ordered subfamily $\mathcal{F}_0$. The union C' of all the sets in $\mathcal{F}_0$ is again a convex set containing C but disjoint from D.

Since C' is maximal, for every $x \notin C'$ we have

$$[x \cup C'] \cap D \neq \emptyset.$$

We will show that, for every $x \notin C'$,

$$C' \cap [x \cup D] = \emptyset.$$

Assume on the contrary that, for some $x \notin C'$, there exists a point $c' \in C' \cap [x \cup D]$ and let $d'' \in [x \cup C'] \cap D$. By Proposition 4 we have $c' \in [x,d']$ for some $d' \in D$ and $d'' \in [x,c'']$ for some $c'' \in C'$. Hence, by **(P)**,

$$[c'',c'] \cap [d',d''] \neq \emptyset.$$

Since $[c'',c'] \subseteq C'$ and $[d',d''] \subseteq D$, this is a contradiction.

Consider now the family $\mathcal{G}$ of all convex sets D'' which contain D but are disjoint from C'. Then $\mathcal{G}$ is nonempty and contains a maximal totally ordered subfamily $\mathcal{G}_0$. The union D' of all the sets in $\mathcal{G}_0$ is again a convex set containing D but disjoint from C'. In the same way, for every $y \notin D'$ we have

$$[y \cup D'] \cap C' \neq \emptyset$$

and

$$D' \cap [y \cup C'] = \emptyset.$$

Since C' is maximal, it follows that $y \in C'$. That is, C' is the complement of D'. □

The axiom **(P)** is essential for the validity of Proposition 22. For suppose $z_1 \in [x,y_1]$, $z_2 \in [x,y_2]$ and $[y_1,z_2] \cap [y_2,z_1] = \emptyset$. If C' and D' are convex sets such that $[y_1,z_2] \subseteq C'$ and $[y_2,z_1] \subseteq D'$, then $x \in C'$ implies $z_1 \in C' \cap D'$ and $x \in D'$ implies $z_2 \in C' \cap D'$.

A set $H \subseteq X$ is said to be a *hemispace* if both H and $X \setminus H$ are convex. Clearly X itself is a hemispace, and the complement of a hemispace is again a hemispace. Sometimes another terminology is more convenient: two sets C,D are said to be a *convex partition* of X if they are convex, nonempty and

$$C \cup D = X, \quad C \cap D = \emptyset.$$

From Proposition 22 we can obtain a separation theorem for any finite number of sets. The proof will be based on the following preliminary result:

LEMMA 23 *Suppose the convex geometry X satisfies* **(P)**. *Let $A = \{a_1,\ldots,a_n\}$ and $B = \{b_1,\ldots,b_n\}$ be subsets of X such that $b_i \in [p,a_i]$ $(i = 1,\ldots,n)$ for some $p \in X$. If $A_i = A \setminus a_i$ and $B_i = B \setminus b_i$, then $\bigcap_{i=1}^n [A_i \cup b_i] \neq \emptyset$, $\bigcap_{i=1}^n [B_i \cup a_i] \neq \emptyset$.*

Proof For $n = 1$ the result is trivial. For $n = 2$ both formulae say that $[a_1,b_2] \cap [a_2,b_1] \neq \emptyset$, and thus the result holds by **(P)**. Suppose now that the result holds for sets of $n \geq 2$ points, and that the hypotheses are satisfied by $A' = A \cup a_{n+1}$ and $B' = B \cup b_{n+1}$. If $x \in \bigcap_{i=1}^n [A_i \cup b_i]$ then

$$[x,a_{n+1}] \subseteq \bigcap_{i=1}^n [A'_i \cup b_i].$$

Since $A'_{n+1} = A$, this establishes the first formula unless $[x,a_{n+1}] \cap [A \cup b_{n+1}] = \emptyset$. In the latter case there exists, by Proposition 22, a convex partition C,D of X such that $[A \cup b_{n+1}] \subseteq C$ and $[x,a_{n+1}] \subseteq D$. Since $p \in D$ implies $b_{n+1} \in D$, we must have $p \in C$. Then $b_n \in C$ and $x \in [A_n \cup b_n] \subseteq C$, which is a contradiction.

The second formula is proved similarly by taking a point $y \in \bigcap_{i=1}^n [B_i \cup a_i]$ and showing that $[y,b_{n+1}] \cap [B \cup a_{n+1}] = \emptyset$ leads to a contradiction. □

PROPOSITION 24 *Suppose the convex geometry X satisfies* **(P)**. *If $C_1,\ldots,C_n$ are convex subsets of X such that $\bigcap_{i=1}^n C_i = \emptyset$, then there exist hemispaces $H_1,\ldots,H_n$ such that $C_i \subseteq H_i$ $(i = 1,\ldots,n)$, $\bigcap_{i=1}^n H_i = \emptyset$ and $\bigcup_{i=1}^n H_i = X$.*

Proof Let $H_1,\ldots,H_n$ be maximal convex sets such that $C_i \subseteq H_i$ $(i = 1,\ldots,n)$, and $\bigcap_{i=1}^n H_i = \emptyset$. Thus any convex set properly containing H_j intersects $\bigcap_{i\neq j} H_i$ $(j = 1,\ldots,n)$. Since there exists a hemispace containing H_j and disjoint from $\bigcap_{i\neq j} H_i$, by Proposition 22, H_j must itself be a hemispace. Assume there exists a point $p \in X \setminus \bigcup_{j=1}^n H_j$. Then $[H_j \cup p]$ intersects $\bigcap_{i\neq j} H_i$ in some point b_j, and $b_j \in [p,a_j]$ for some $a_j \in H_j$. Applying Lemma 23 to the sets $A = \{a_1,\ldots,a_n\}$ and $B = \{b_1,\ldots,b_n\}$, we see that the sets $[(B \setminus b_j) \cup a_j]$ $(j = 1,\ldots,n)$ have a common point. Since $(B \setminus b_j) \cup a_j \subseteq H_j$, this contradicts $\bigcap_{j=1}^n H_j = \emptyset$. □

Finally we consider some ways of manufacturing new convex geometries from given ones.

(i) *Restriction*: Let X be a convex geometry and Y a nonempty convex subset of X. Since $x,y \in Y$ implies $[x,y] \subseteq Y$, by restriction to Y we again obtain a convex geometry. Moreover this convex geometry on Y satisfies those of the axioms **(L2)**,**(L3)**,**(L4)**,**(P)** which are satisfied by the given convex geometry on X.

(ii) *Join*: Let $X = X_1 \cup X_2$, where X_1,X_2 are disjoint sets on each of which a convex geometry is defined. For $a,b \in X$ we define $[a,b] = \{a,b\}$ if $a \in X_1$, $b \in X_2$ or if $a \in X_2$, $b \in X_1$, and $[a,b] = [a,b]_k$ if $a,b \in X_k$ $(k = 1,2)$. It is easily verified that this defines a convex geometry on X. Indeed the axiom **(C)** holds trivially if $c \in \{a_1,b_1\}$ or if $d \in \{c,b_2\}$ and otherwise holds in X because it holds in X_1 and X_2. Moreover this convex geometry on X satisfies those of the axioms **(L2)**,**(L3)**,**(L4)**,**(P)** which are satisfied by both the given convex geometries on X_1,X_2.

4 EXAMPLES (continued)

It will now be shown that a real vector space, or indeed a vector space over any ordered division ring, is a convex geometry with the usual definition of segments. Since **(L1)** is obviously satisfied, we need only prove **(C)**. The geometric significance of the axiom **(C)** in this case is illustrated in Figure 3. We have $d = \theta c + (1-\theta)b_2$ and $c = \lambda a + (1-\lambda)b_1$, where $0 \le \theta,\lambda \le 1$. Evidently

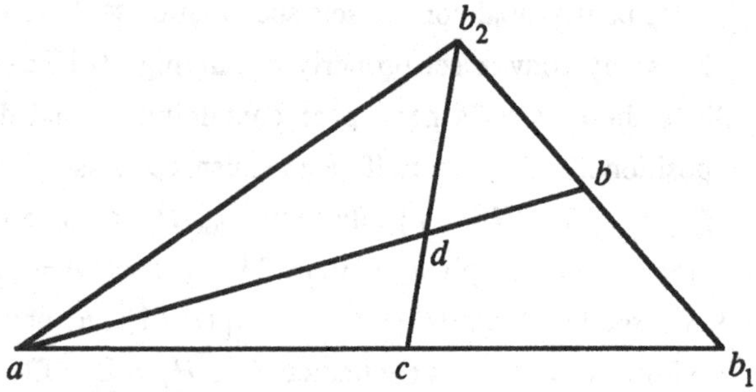

Figure 3

we may assume that $0 < \theta, \lambda < 1$. Hence if we put $\nu = 1 - \theta\lambda$, then $0 < \nu < 1$. Moreover $d = (1 - \nu)a + \nu b$, where

$$b = \nu^{-1}\theta(1 - \lambda)b_1 + \nu^{-1}(1 - \theta)b_2 \in [b_1, b_2].$$

We are going to show that segments in a vector space over an ordered division ring also satisfy the axioms **(L2)**–**(L4)** and **(P)**. It is easy to verify **(L2)**; for suppose

$$b = \lambda c + (1 - \lambda)a, \quad c = \mu b + (1 - \mu)d,$$

where $0 \leq \lambda, \mu < 1$. Substituting the second equation in the first, we obtain $b = \theta a + (1 - \theta)d$, where $\theta = (1 - \lambda\mu)^{-1}(1 - \lambda)$ and thus $0 < \theta \leq 1$. Thus a vector space over an ordered division ring is actually an aligned space with the anti-exchange property.

We can now give a concrete illustration of the results of Chapter I, which was first proved by Doignon (1973):

PROPOSITION 25 *Let $\mathbf{Z}^d$ denote the set of points in $\mathbb{R}^d$ whose coordinates are all integers. If S is any finite subset of $\mathbf{Z}^d$ with $|S| > 2^d$, then $\bigcap_{s \in S} [S \setminus s]$ contains a point of $\mathbf{Z}^d$.*

Proof Put $X = [S] \cap \mathbf{Z}^d$. Then $[X] = [S]$, since $S \subseteq X \subseteq [S]$, and hence $X = [X] \cap \mathbf{Z}^d$. As we saw in Chapter I, X becomes an aligned space with the anti-exchange property if we define a 'convex' set in X to be the intersection with X of a convex set in $\mathbb{R}^d$. We wish to show that S is not a Helly set in this aligned space.

Assume that S is a Helly set. Then, by Proposition I.15, there exists a Helly set $T \subseteq X$ with $|T| = |S|$ and $T = [T] \cap X$. Since $|T| > 2^d$, there exist points

$y,z \in T$ whose corresponding coordinates are all congruent (mod 2). Then $x = (y+z)/2 \in \mathbf{Z}^d$. But $x \in [T] \subseteq [X] = [S]$. Hence $x \in X$, and so $x \in T$. Thus T is not independent, which contradicts Proposition I.9. □

Since the set S of all points in $\mathbf{Z}^d$ whose coordinates are either 0 or 1 is an independent set such that $S = [S] \cap \mathbf{Z}^d$, it follows from the remark after the proof of Proposition I.15 that the bound 2^d in Proposition 25 is actually sharp.

It is obvious that the axioms **(L3)** and **(L4)** are satisfied in a vector space over an ordered division ring with the usual definition of segments. Finally it will be shown that the axiom **(P)** is also satisfied. The geometric significance of the axiom **(P)** in this case is illustrated in Figure 4.

Suppose

$$c_1 = \lambda b_1 + (1-\lambda)a, \quad c_2 = \mu b_2 + (1-\mu)a$$

for some λ,μ with $0 \le \lambda,\mu \le 1$. If $\lambda = \mu = 1$ then **(P)** obviously holds, since $c_1 = b_1$ and $c_2 = b_2$. Otherwise we have $\lambda\mu < 1$ and $\mu\lambda < 1$. Thus if we put

$$\alpha = (1-\lambda)(1-\mu\lambda)^{-1}, \quad \beta = (1-\mu)(1-\lambda\mu)^{-1},$$

then $0 \le \alpha,\beta \le 1$. Moreover, since $(1-\lambda\mu)^{-1} = \lambda(1-\mu\lambda)^{-1}\lambda^{-1}$, we have

$$(1-\alpha)b_1 + \alpha c_2 = (1-\beta)b_2 + \beta c_1.$$

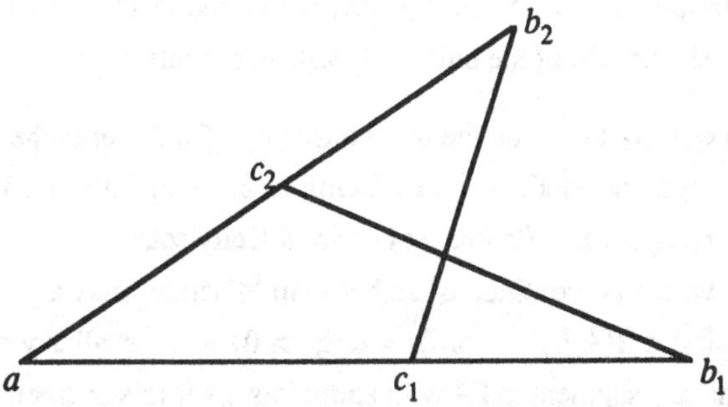

Figure 4

We leave it as an exercise for the interested reader to verify that Examples 4–8 at the beginning of this chapter are all convex geometries. Furthermore, Examples 4–7 satisfy the axiom **(L2)**. The axioms **(L3)**,**(L4)** and **(P)** are in general not satisfied in Example 4, but they are all satisfied if the set X is totally ordered. The axioms (L4) and (P) are satisfied in Examples 5–6, but the axiom (L3) is in general not satisfied. The axiom (P) is satisfied in Examples 7–8. However, the axioms (L3)–(L4) are not satisfied in Example 7, and the axioms (L2)–(L4) are not satisfied in Example 8.

Finally we show that the axioms studied in this chapter are mutually independent.

EXAMPLE 10 We exhibit systems which satisfy all the axioms (C),(L1)–(L4),(P) with the exception of the one named:

(C) Take $X = \{a,b_1,b_2,c,d\}$ with $[a,b_1] = \{a,b_1,c\}$, $[c,b_2] = \{c,b_2,d\}$, and $[x,y] = \{x,y\}$ otherwise. The collection of all sets $C \subseteq X$ such that $c \in C$ if $a,b_1 \in C$, and $d \in C$ if $c,b_2 \in C$, is an alignment on X and, furthermore, it has the anti-exchange property. Proposition 7 no longer holds, since $\{a,b_1,b_2\}$ is a Carathéodory set but $\{b_1,b_2\}$ is not.

(L1) Take any set X such that $|X| > 1$ with $[x,y] = X$ for all $x,y \in X$.

(L2) Take $X = \{a;b,c,d\}$ with $[a,c] = \{a,b,c\}$, $[b,d] = \{b,c,d\}$ and $[x,y] = \{x,y\}$ otherwise.

An example which is 'continuous' rather than 'discrete' may also be given. Let $X = \{z = e^{i\theta} \in \mathbb{C}: 0 \leq \theta < \pi/3$ or $2\pi/3 \leq \theta < \pi$ or $4\pi/3 \leq \theta < 5\pi/3\}$. No two points of X are antipodal and, for all $x,y \in X$, we define $[x,y]$ to be the set of all points of X on the shorter arc of the unit circle with endpoints x,y.

(L3) Take $X = \{x_0,x_1,x_2,x_3\}$ to be the tree consisting of three branches joining x_0 to x_1,x_2,x_3, with segments defined as in Example 6. Proposition 13 no longer holds, since $S = \{x_1,x_2,x_3\}$ is a Radon set but not a Helly set.

An example which is 'continuous' rather than 'discrete' may again be given. Let $X = \{x = (\xi_1,\xi_2) \in \mathbb{R}^2: \xi_2 = 0$ or $\xi_1 = 0, \xi_2 > 0\}$ and for all $x,y \in X$ define $[x,y]$ to be the ordinary segment in $\mathbb{R}^2$ with endpoints x,y if this segment is entirely contained in X, and otherwise to be the union of the ordinary segments $[0,x]$ and $[0,y]$, where $0 = (0,0)$.

(L4) Take $X = \{a,b,c,d\}$ with $[a,b] = X$ and $[x,y] = \{x,y\}$ otherwise. Proposition 14 no longer holds, since $[a,b] \neq [a,c] \cup [b,c]$.

(P) Take $X = \{a,a',b,b',c\}$ with $[a,c] = \{a,a',c\}$, $[b,c] = \{b,b',c\}$, and $[x,y] = \{x,y\}$ otherwise.

5 NOTES

Examples 4–6 are studied as anti-exchange alignments, rather than as convex geometries, by Jamison-Waldner (1982). Examples 7 and 8 appear in Prenowitz (1961) and van de Vel (1993) respectively.

The earliest paper on axiomatic convexity is perhaps Levi (1951), who used the conclusion of Proposition 14 as one of his axioms and also proved Proposition I.9. Another early paper is Ellis (1952), where Proposition 22 is proved under equivalent hypotheses to those used here. Our approach is closer to that of Prenowitz (1961) and Calder (1971).

The books of Soltan (1984) and van de Vel (1993) are devoted to the axiomatic theory of convexity and have extensive bibliographies. The proof of Proposition 24 is taken from van de Vel. In these works there is considerable freedom in the choice of axioms. Our choice of axioms is directed by the wish to obtain ultimately a set of axioms for Euclidean geometry.

III
Linearity

In this chapter we show how *lines* may be defined, and how the points of a line may be totally ordered, using four of the axioms considered in Chapter II. The notion of a *linear geometry*, which is defined by all six of the axioms of Chapter II, is then introduced. A linear geometry possesses both an anti-exchange alignment of *convex* sets and an exchange alignment of *affine* sets. This rich structure makes it possible to establish many familiar geometrical properties, including the existence of *half-spaces* associated with a *hyperplane* and the property which Pasch showed to have been omitted by Euclid. Nevertheless a linear geometry may in fact possess only finitely many points. However, it is shown that all *irreducible* linear geometries which are not dense, in the sense of Chapter V, have a rather simple form.

In conclusion we establish the theorems of Helly, Radon and Carathéodory in any linear geometry. The extension to an arbitrary linear geometry of the theorems of Bárány and Tverberg, which we prove for $\mathbb{R}^d$, is left as an open problem.

1 LINES

Let X be a set and suppose that with any unordered pair $\{a,b\}$ of elements of X there is associated a subset $[a,b]$ of X containing a and b. The elements of X will be called *points* and the subsets $[a,b]$ *segments*. *Throughout this section we will assume that the following four axioms are satisfied:*

(L1) $[a,a] = \{a\}$;
(L2) *if* $b \in [a,c]$, $c \in [b,d]$ *and* $b \neq c$, *then* $b \in [a,d]$;
(L3) *if* $c \notin [a,b]$ *and* $b \notin [a,c]$, *then* $[a,b] \cap [a,c] = \{a\}$;
(L4) *if* $c \in [a,b]$, *then* $[a,b] = [a,c] \cup [c,b]$.

We propose to study the consequences of this assumption.

PROPOSITION 1 *For all* $a,b,c,d \in X$,

(i) *if* $c \in [a,b]$ *and* $d \in [a,c]$, *then* $c \in [b,d]$;
(ii) *if* $c \in [a,b]$ *and* $b \in [a,c]$, *then* $b = c$;
(iii) *if* $c \in [a,b]$, *then* $[a,c] \cap [b,c] = \{c\}$.

Proof (ii) Assume $b \neq c$. Then $b \in [a,a]$, by **(L2)**, and hence $b = a$, by **(L1)**. Thus $c \in [a,a]$ and hence $c = a$. Then $b = c$, which is a contradiction.

(iii) Suppose $d \in [a,c] \cap [b,c]$. Then $d \in [a,b]$, by **(L4)**. By **(L4)** also, either $c \in [a,d]$ or $c \in [b,d]$. In either event $d = c$, by (ii).

(i) Obviously we may assume $d \neq a,c$. Then $c \notin [a,d]$, by (ii). By **(L4)**, $d \in [a,b]$. Since $c \notin [a,d]$, it follows from **(L4)** again that $c \in [b,d]$. □

We recall that Proposition 1 was already proved, under different hypotheses, in Proposition II.18.

If a and b are distinct points, we define the *line* $\langle a,b \rangle$ to be the set of all points c such that either $c \in [a,b]$ or $a \in [b,c]$ or $b \in [c,a]$. If $a = b$ we set $\langle a,b \rangle = \{a\}$.

It is easily verified that if X is a vector space over any ordered division ring D, with the standard definition of segments $[a,b]$, then the line $\langle a,b \rangle$ is the set of all $c \in X$ which can be represented in the form $c = \lambda a + (1 - \lambda)b$, where $\lambda \in D$.

Clearly $\langle a,b \rangle = \langle b,a \rangle$ and $[a,b] \subseteq \langle a,b \rangle$. In particular, $a,b \in \langle a,b \rangle$. Furthermore, if a,b,c are distinct points such that $c \in \langle a,b \rangle$, then also $a \in \langle b,c \rangle$ and $b \in \langle c,a \rangle$.

PROPOSITION 2 *If* $c,d \in \langle a,b \rangle$ *and* $c \neq d$, *then* $\langle c,d \rangle = \langle a,b \rangle$.

Proof Clearly we must have $a \neq b$. It is sufficient to show that if $c \neq a,b$ then $\langle a,c \rangle = \langle a,b \rangle$. In fact, by symmetry we need only show that $\langle a,c \rangle \subseteq \langle a,b \rangle$. The following proof is arranged so as to appeal to Proposition 1(i) rather than **(L4)**, as far as possible.

Let $x \in \langle a,c \rangle$, so that either $x \in [a,c]$ or $a \in [c,x]$ or $c \in [a,x]$. We wish to show that either $x \in [a,b]$ or $a \in [b,x]$ or $b \in [a,x]$. Evidently we may assume that $x \neq a,b,c$.

Suppose first that $c \in [a,b]$. If $a \in [c,x]$ then $a \in [b,x]$, by **(L2)**. If $c \in [a,x]$ then $x \in [a,b]$ or $b \in [a,x]$, by **(L3)**. If $x \in [a,c]$ then $c \in [b,x]$, by Proposition 1(i), and hence $x \in [a,b]$, by **(L2)**.

Suppose next that $a \in [b,c]$. If $x \in [a,c]$ then $a \in [b,x]$, by Proposition 1(i). If $a \in [c,x]$ then $b \in [c,x]$ or $x \in [b,c]$, by **(L3)**. Moreover, by Proposition 1(i), $b \in [c,x]$ implies $b \in [a,x]$ and $x \in [b,c]$ implies $x \in [a,b]$. If $c \in [a,x]$ then $a \in [b,x]$, by **(L2)**.

Suppose finally that $b \in [a,c]$. If $a \in [c,x]$ then $a \in [b,x]$, by Proposition 1(i). If $c \in [a,x]$ then $c \in [b,x]$, by Proposition 1(i), and hence $b \in [a,x]$, by **(L2)**. If $x \in [a,c]$ then $x \in [a,b]$ or $x \in [b,c]$, by **(L4)**. Moreover, by Proposition 1(i), $x \in [b,c]$ implies $b \in [a,x]$. □

COROLLARY 3 *If there exist three distinct points a,b,c such that* $c \notin \langle a,b\rangle$, *then for any distinct points x,y there exists a point z such that* $z \notin \langle x,y\rangle$. □

Points which lie on the same line will be said to be *collinear.* We show next how the points of each line may be totally ordered:

PROPOSITION 4 *Given two distinct points* $x,y \in X$, *the points of the line* $\ell = \langle x,y\rangle$ *may be totally ordered so that* $x \le y$ *and so that, for any points* $a,b \in \ell$ *with* $a \le b$, *the segment* $[a,b]$ *consists of all* $c \in \ell$ *such that* $a \le c \le b$. *Moreover, this total ordering is unique.*

Proof For any $a,b \in \ell$, we write $a \le b$ if either of the following conditions is satisfied:

(i) $a \in [x,b]$ and either $y \in [x,b]$ or $b \in [x,y]$,

(ii) $x \in [a,y]$ and either $b \in [a,y]$ or $y \in [a,b]$.

It should be noted that (i) and (ii) say the same thing if $a = x$ and that they cannot both hold if $a \neq x$.

The definition obviously implies $x \le y$. Also, from the definition of a line it is clear that for any $a \in \ell$ we have $a \le a$. Let a,b be points of ℓ such that both $a \le b$ and $b \le a$. We wish to show that this implies $b = a$.

Assume first that $a \in [x,b]$ and either $y \in [x,b]$ or $b \in [x,y]$. Since $x \neq y$, this implies $x \notin [b,y]$. Since $b \le a$, it follows that $b \in [x,a]$. Hence $b = a$, by Proposition 1(ii). The case $b \in [x,a]$ and either $y \in [x,a]$ or $a \in [x,y]$ may be discussed similarly. Suppose finally that $x \in [a,y]$ and $x \in [b,y]$. Then either

$a \in [b,y]$ or $b \in [a,y]$, by **(L3)**. We cannot have also $y \in [a,b]$, since this would imply $y = a$ or $y = b$, respectively, and hence $x = y$. Consequently we must have $b \in [a,y]$ and $a \in [b,y]$, which implies $b = a$.

Suppose next that a,b,c are distinct points of ℓ such that $a \leq b$ and $b \leq c$. We wish to show that this implies $a \leq c$.

Assume first that $b \in [x,c]$ and either $y \in [x,c]$ or $c \in [x,y]$. If $a \in [x,b]$ then $a \in [x,c]$ and hence $a \leq c$. If $x \in [a,y]$ then $y \in [x,c]$ implies $y \in [a,c]$, by **(L2)**, and $c \in [x,y]$ implies $c \in [a,y]$, by **(L4)**. In both cases $a \leq c$. Assume next that $x \in [b,y]$ and either $c \in [b,y]$ or $y \in [b,c]$. If $x \in [a,y]$ and $b \in [a,y]$ then $c \in [a,y]$ or $y \in [a,c]$, respectively. Thus again $a \leq c$. All remaining cases may be discussed similarly.

It remains to show that if a,b are distinct points of ℓ then either $a \leq b$ or $b \leq a$. Assume that neither relation holds. Then if $a \in [x,b]$ we must have $x \in [b,y]$, and if $x \in [b,y]$ we must have $b \in [a,y]$. Hence $b \in [x,y]$, by **(L2)**, and $b = x$, by Proposition 1(ii). But this implies $b = a$, which contradicts our assumption. Therefore $a \notin [x,b]$. Similarly we can show that $b \notin [a,x]$. Consequently $x \in [a,b]$.

If $x \in [a,y]$ we must have $a \in [b,y]$ and hence $x \in [b,y]$, and if $x \in [b,y]$ we must have $b \in [a,y]$. Therefore $x \notin [a,y]$, and similarly $x \notin [b,y]$. Thus either $a \in [x,y]$ or $y \in [a,x]$. But $a \in [x,y]$ would imply $x \in [b,y]$, by **(L2)**, and $y \in [a,x]$ would imply $x \in [b,y]$, by Proposition 1(i). Thus we cannot escape a contradiction.

This proves that the relation $\leq$ is a total ordering of the line ℓ such that $x \leq y$. Suppose now that $a \leq b$ and $c \in [a,b]$. If (i) holds, then $c \in [x,b]$ and hence $c \leq b$. Moreover $a \in [x,c]$ and either $y \in [x,c]$ or $c \in [x,y]$, so that $a \leq c$. Similarly it may be seen that $a \leq c \leq b$ if (ii) holds.

Suppose on the other hand that $c \in \ell$ and $a \leq c \leq b$. We will show that $c \in [a,b]$. Indeed, if $a \in [c,b]$ then, by what we have just proved, $c \leq a \leq b$ and hence $c = a$. Similarly if $b \in [a,c]$, then $c = b$.

Finally, suppose that $\leq$ is any total ordering of ℓ with the properties in the statement of the proposition. Let a,b be any distinct elements of ℓ and assume first that (i) holds. If $b \leq x$, then $b \leq a \leq x$ and $x \leq b \leq y$, since $y \notin [x,b]$. Hence $b = a$, which is a contradiction. Consequently $x \leq b$ and $x \leq a \leq b$. Assume next that (ii) holds. Then $a \leq x \leq y$ and either $a \leq b \leq y$ or $a \leq y \leq b$, since $y \neq a$. In any event $a \leq b$, and so the total ordering is the one originally defined. □

To illustrate the application of Proposition 4 we prove

PROPOSITION 5 *The union of two segments with more than one common point is again a segment.*

Proof Let $[a,b]$ and $[c,d]$ be two segments whose intersection contains the distinct points x,y. Assume the line $\ell = \langle x,y \rangle$ totally ordered, as in Proposition 4, and choose the notation so that $a \leq b$, $c \leq d$ and $a \leq c$. Then $y \leq b$, since $y \in [a,b]$, and $c \leq x$, since $x \in [c,d]$. If $d \leq b$, then $[c,d] \subseteq [a,b]$. On the other hand if $b \leq d$, then $a \leq c \leq x \leq y \leq b \leq d$ and hence

$$[a,b] \cup [c,d] = [a,c] \cup [c,d] = [a,d]. \quad \square$$

The following result shows that if all points of X are collinear, then the axioms **(C)** and **(P)** of Chapter II both hold, and actually in a stronger form.

PROPOSITION 6 *Suppose a,b_1,b_2 are collinear points and $c_1 \in [a,b_1]$, $c_2 \in [a,b_2]$. Then either $c_1 \in [b_1,c_2]$ or $c_2 \in [b_2,c_1]$. Moreover, if $d \in [c_1,c_2]$ then either $d \in [a,b_1]$ or $d \in [a,b_2]$.*

Proof Obviously we may assume $c_1 \neq c_2$. If $c_2 \in [a,c_1]$ then $c_1 \in [b_1,c_2]$, by Proposition 1(i). Similarly if $c_1 \in [a,c_2]$ then $c_2 \in [b_2,c_1]$. Hence we may suppose $a \in (c_1,c_2)$. But then both $c_1 \in [b_1,c_2]$ and $c_2 \in [b_2,c_1]$, by **(L2)**.

If $a \in [b_1,b_2]$, then $c_1,c_2 \in [b_1,b_2]$ and hence $d \in [b_1,b_2]$. Thus $d \in [a,b_1]$ or $d \in [a,b_2]$, by **(L4)**. If $b_1 \in [a,b_2]$, then $c_1 \in [a,b_2]$ and hence $d \in [a,b_2]$. Similarly if $b_2 \in [a,b_1]$, then $c_2 \in [a,b_1]$ and hence $d \in [a,b_1]$. $\square$

The concept of collinearity makes it possible to obtain sharper forms (under stronger hypotheses) for a number of results in Chapter II. These sharper forms are based on the following simple results:

(a) *Let a,b_1,b_2 be non-collinear points and suppose $c_1 \in (a,b_1)$, $c_2 \in (a,b_2)$. If $d \in [b_1,c_2] \cap [b_2,c_1]$, then actually $d \in (b_1,c_2) \cap (b_2,c_1)$.*

Proof It is sufficient to show that $d \neq b_1,c_1$. By Proposition 2 $d = b_1$ implies $b_2 \in \langle b_1,c_1 \rangle = \langle a,b_1 \rangle$, which is a contradiction. Similarly $d = c_1$ implies $c_2 \in \langle a,b_1 \rangle$ and again $b_2 \in \langle a,b_1 \rangle$. $\square$

Similarly we can prove

(b) *Let a,b_1,b_2 be non-collinear points and suppose $c_1 \in (a,b_1)$, $c_2 \in (a,b_2]$, $c \in (c_1,c_2)$. If $c \in [a,b]$ for some $b \in [b_1,b_2]$, then actually $c \in (a,b)$ and $b \in (b_1,b_2)$.* □

(c) *Let a,b_1,b_2 be non-collinear points and suppose $c_1 \in (a,b_1)$, $c_2 \in (a,b_2]$, $b \in (b_1,b_2)$. If there exists a point $c \in [a,b] \cap [c_1,c_2]$, then actually $c \in (a,b) \cap (c_1,c_2)$.* □

(d) *Let a,a',b,b' be non-collinear points and let $x \in (a,a') \cap (b,b')$. If $x \in [y,y']$ for some $y \in (a,b)$ and $y' \in [a',b']$, then actually $x \in (y,y')$ and $y' \in (a',b')$.* □

By combining (a) with **(P)**, and (b),(c),(d) with Propositions II.2,17,19 respectively, we obtain, *for convex geometries satisfying the additional axioms* **(L2)–(L4)** *and* **(P)**:

(P)′ *if a,b_1,b_2 are non-collinear points and if $c_1 \in (a,b_1)$, $c_2 \in (a,b_2)$, then there exists a point $d \in (b_1,c_2) \cap (b_2,c_1)$.* □

PROPOSITION II.2′ *Let a,b_1,b_2 be non-collinear points and let $c_1 \in (a,b_1)$, $c_2 \in (a,b_2]$. If $c \in (c_1,c_2)$, then $c \in (a,b)$ for some $b \in (b_1,b_2)$.* □

PROPOSITION II.17′ *Let a,b_1,b_2 be non-collinear points and let $c_1 \in (a,b_1)$, $c_2 \in (a,b_2]$. If $b \in (b_1,b_2)$, then there exists a point $c \in (a,b) \cap (c_1,c_2)$.* □

PROPOSITION II.19′ *Let a,a',b,b' be non-collinear points. If $y \in (a,b)$ and $x \in (a,a') \cap (b,b')$, then $x \in (y,y')$ for some $y' \in (a',b')$.* □

2 LINEAR GEOMETRIES

We define a *linear geometry* to be a convex geometry satisfying the additional axioms **(L2)–(L4)** and **(P)**. An example of a linear geometry with only finitely many points is the set $X = \{a,a',b,b',c\}$ with $[a,a'] = \{a,a',c\}$, $[b,b'] = \{b,b',c\}$ and $[x,y] = \{x,y\}$ otherwise. In spite of such examples we are going to show that linear geometries possess many of the properties of real vector spaces.

Throughout the remainder of this chapter we will assume that a set X is given on which a linear geometry is defined. We define a subset A of X to be *affine* if $x,y \in A$ implies $<x,y> \subseteq A$. For example, the line $<x,y>$ is itself affine, by Proposition 2.

It follows at once from the definition that the collection $\mathcal{A}$ of all affine sets is a normed alignment on X. For any set $S \subseteq X$, we define its *affine hull* $<S>$ to be the intersection of all affine sets which contain S. Our notations are consistent, since the affine hull of the set $\{x,y\}$ is the line $<x,y>$. Evidently, for any set $S \subseteq X$, we have $S \subseteq [S] \subseteq <S>$. The next result, inspired by Bennett and Birkhoff (1985), is of fundamental importance:

PROPOSITION 7 *If C is a convex set, then*

$$<C> = \bigcup_{x,x' \in C} <x,x'>.$$

Proof It is sufficient to show that the set $A = \bigcup_{x,x' \in C} <x,x'>$ is affine, since it is certainly contained in any affine set which contains C. Evidently we may assume that C contains more than one point. Thus we wish to show that if $b \in <x_1,x_2>$, $c \in <x_3,x_4>$, and $a \in <b,c>$, where $x_1,...,x_4 \in C$ and $x_1 \neq x_2$, $x_3 \neq x_4$, $b \neq c$, then there exist $x_5,x_6 \in C$ with $x_5 \neq x_6$ such that $a \in <x_5,x_6>$. This is evident if $c \in <x_1,x_2>$, since then also $a \in <x_1,x_2>$. Consequently we assume $c \notin <x_1,x_2>$, and similarly $b \notin <x_3,x_4>$.

By symmetry, and the convexity of C, we need only consider the following five cases:

(A) $x_2 \in [b,x_1]$, $x_4 \in [c,x_3]$, $c \in [b,a]$;
(B) $x_2 \in [b,x_1]$, $x_4 \in [c,x_3]$, $a \in [b,c]$;
(C) $x_2 \in [b,x_1]$, $c \in [x_3,x_4]$, $a \in [b,c]$;
(D) $x_2 \in [b,x_1]$, $c \in [x_3,x_4]$, $c \in [a,b]$;
(E) $x_2 \in [b,x_1]$, $c \in [x_3,x_4]$, $b \in [a,c]$.

Consider first case (A). By **(P)** there exists a point $y \in [a,x_2] \cap [c,x_1]$. Since $c \notin <x_1,x_2>$, we have $y \neq x_2$ and hence $a \in <x_2,y>$. Consequently we may assume $y \notin C$. By **(P)** also there exists a point $x_5 \in [y,x_3] \cap [x_1,x_4]$. Then $x_5 \in C$ and $x_5 \in [a,x_2,x_3]$. Hence $x_5 \in [a,x_6]$ for some $x_6 \in [x_2,x_3]$. Thus $x_6 \in C$ and $a \in <x_5,x_6>$ if $x_5 \neq x_6$. We can now assume $x_5 = x_6 \in [x_2,x_3]$. If $x_3 \neq x_5$ then $x_2 \in <x_3,x_5>$, $y \in <x_3,x_5>$, and hence $a \in <x_3,x_5>$. Consequently we can now assume $x_3 = x_5 \in [x_1,x_4]$. Then, by **(L2)**, $x_4 \in [c,x_1]$. Since $y \notin C$ it follows from **(L4)** that $y \in [c,x_4]$ and then from Proposition 1(i) that

$x_4 \in [x_1,y] \subseteq [a,x_1,x_2]$. Hence $x_4 \in [a,x_7]$ for some $x_7 \in [x_1,x_2]$. Then $x_7 \in C$ and $a \in \langle x_4,x_7 \rangle$ if $x_4 \neq x_7$. In fact we cannot have $x_4 = x_7$, since this would imply $x_4 \in [x_1,x_2]$, $x_3 \in [x_1,x_2]$ and hence

$$c \in \langle x_3,x_4 \rangle \subseteq \langle x_1,x_2 \rangle.$$

Similarly in case (B) there exist points $y \in [b,x_4] \cap [a,x_3]$ and $x_5 \in [y,x_1] \cap [x_2,x_4]$. The argument can now be completed as in the previous case.

In case (C) there exists a point $x_5 \in [c,x_2] \cap [a,x_1]$. Moreover $x_5 \in C$, since $c \in C$, and $x_5 \neq x_1$, since $c \notin \langle x_1,x_2 \rangle$. Hence $a \in \langle x_1,x_5 \rangle$.

Similarly in case (D) there exists a point $x_5 \in [a,x_2] \cap [c,x_1]$ and the argument can be completed as in the previous case.

Finally in case (E) we have $c \in C$ and $x_2 \in [a,c,x_1]$. Hence $x_2 \in [a,x_5]$ for some $x_5 \in [c,x_1]$. Then $x_5 \in C$, $x_5 \neq x_2$ and $a \in \langle x_2,x_5 \rangle$. □

The definitions and results of Chapter I, Section 2 may be applied to the alignment of affine sets on X, as well as to the alignment of convex sets on X. To avoid confusion, the qualification 'affine' will be made explicit. Thus we say that a set $S \subseteq X$ is *affine independent* if, for every $x \in S$, $x \notin \langle S \setminus x \rangle$. A subset T of a set S is an *affine generator* of S if $\langle T \rangle = \langle S \rangle$ and an *affine basis* of S if, in addition, T is affine independent.

According to the results of Chapter I, Section 2, an infinite set is affine independent if every finite subset is affine independent, and any affine independent set is contained in a maximal affine independent set. It will now be shown that the alignment of affine sets is an *exchange* alignment:

PROPOSITION 8 *For any set $S \subseteq X$, if $y \in \langle S \cup x \rangle$ but $y \notin \langle S \rangle$, then $x \in \langle S \cup y \rangle$.*

Proof Obviously we may assume that $x \notin [S \cup y]$. By Proposition 7, $y \in \langle z_1,z_2 \rangle$, where $z_1,z_2 \in [S \cup x]$. Therefore, by Proposition II.4, $z_1 \in [x,w_1]$ and $z_2 \in [x,w_2]$ for some $w_1,w_2 \in [S]$.

If $y \in [z_1,z_2]$, then $y \in [x,w_1,w_2]$ and hence $y \in [x,w]$ for some $w \in [w_1,w_2] \subseteq [S]$. Thus $y \neq w$ and $x \in \langle y,w \rangle \subseteq \langle S \cup y \rangle$.

By symmetry it only remains to consider the case $z_2 \in (y,z_1)$. We will assume $x \notin \langle y,w_1,w_2 \rangle$ and derive a contradiction.

Since $x \notin \langle y,w_1,w_2\rangle$, we must have $z_1 \neq x,w_1$. Thus $z_1 \in (x,w_1)$ and hence, by Proposition II.2′, $z_2 \in (x,w)$ for some $w \in (y,w_1)$. Hence $w \in \langle x,z_2\rangle = \langle x,w_2\rangle$. Since $x \notin \langle y,w_1,w_2\rangle$, we must have $w = w_2$. Since $y \in \langle w,w_1\rangle$, this is a contradiction. □

It follows that all the results of Chapter I, Section 5 are valid for the alignment of affine sets in a linear geometry. With the definition of *dimension* given there, the empty set ∅ has dimension –1, a point has dimension 0 and a line has dimension 1. An affine set of dimension 2 will be called a *plane*. For finite-dimensional X, an affine set $H \subseteq X$ is a hyperplane if and only if $\dim H = \dim X - 1$.

Pasch (1882) pointed out the incompleteness of Euclid's axioms for geometry and introduced the following additional axiom: 'If a line in the plane of a triangle does not pass through any of its vertices but intersects one of its sides, then it also intersects another of its sides.' It will now be shown that Pasch's axiom holds in any linear geometry. (It is not difficult to show that, conversely, any convex geometry is a linear geometry if it satisfies the axioms **(L2)–(L4)** and Pasch's axiom.)

LEMMA 9 *If a,b,c,d are points such that*

$$[a,b] \cap \langle c,d\rangle = \emptyset, \quad [a,c] \cap \langle b,d\rangle = \emptyset, \quad [b,c] \cap \langle a,d\rangle = \emptyset,$$

then $d \notin \langle a,b,c\rangle$.

Proof The hypotheses evidently imply that a,b,c are not collinear and $d \notin [a,b] \cup [a,c] \cup [b,c]$. We will assume $d \in \langle a,b,c\rangle$ and derive a contradiction.

It follows from Proposition II.16 that if $d \in [a,b,c]$, then $d \in [a,e]$ for some $e \in [b,c]$, which contradicts $[b,c] \cap \langle a,d\rangle = \emptyset$. Thus we now suppose $d \notin [a,b,c]$.

By Proposition 7, $d \in \langle p,q\rangle$, where $p,q \in [a,b,c]$ and $p \neq q$. Moreover we may choose the notation so that $q \in (p,d)$. Furthermore, by Proposition II.16, we may take $p,q \in [a,b] \cup [a,c] \cup [b,c]$.

If $p,q \in [a,b]$, then $d \in \langle a,b\rangle$ and either $a \in (b,d)$ or $b \in (a,d)$. In either case $\langle a,d\rangle \cap [b,c] \neq \emptyset$. It is obvious also that $\langle a,d\rangle \cap [b,c] \neq \emptyset$ if $q = a$ and $p \in [b,c]$ or if $p = a$ and $q \in [b,c]$.

By symmetry it only remains to consider the case $p \in (a,b)$ and $q \in (b,c)$. Then $q \in (b,r)$ for some $r \in (a,d)$. Since $q \in (b,c)$ and $r \neq c$, either $r \in (b,c)$ or $c \in (b,r)$. If $r \in (b,c)$, then $(a,d) \cap (b,c) \neq \varnothing$. If $c \in (b,r)$, then $c \in (q,r)$ and hence $c \in (d,e)$ for some $e \in (a,p) \subseteq (a,b)$. Thus $\langle c,d\rangle \cap (a,b) \neq \varnothing$. □

PROPOSITION 10 *Let a,b,c be non-collinear points and let ℓ be a line in the plane $\langle a,b,c\rangle$ such that $a,b,c \notin \ell$. If ℓ intersects (a,b), then ℓ also intersects either (a,c) or (b,c), but not both.*

Proof We show first that ℓ cannot intersect (a,b), (a,c) and (b,c). By symmetry it is sufficient to show that if $d \in (a,b)$, $e \in (a,c)$ and $f \in (b,c)$, then $f \notin (d,e)$. But if $f \in (d,e)$ then, by Proposition II.2′, $f \in (a,g)$ for some $g \in (b,c)$. Hence $a \in \langle b,c\rangle$, which is a contradiction.

It remains to show that ℓ intersects two 'sides' of the 'triangle' $[a,b,c]$. Since $a,b,c \notin \ell$, this follows at once from Proposition II.16 if ℓ contains two distinct points of $[a,b,c]$. Thus we now assume that ℓ contains a point $e \notin [a,b,c]$. Since $e \in \langle a,b,c\rangle$, it follows from Lemma 9 that the points a,b,c may be named so that there exists a point $p \in \langle a,e\rangle \cap [b,c]$. Thus the hypothesis is now that ℓ contains a point $d \in (a,b) \cup (a,c) \cup (b,c)$.

Suppose first that $d \in (a,b)$, which implies $p \neq b$. If $a \in (e,p)$, then $d \in (e,f)$ for some $f \in (b,p)$. If $p \in (a,e)$, then there exists a point $f \in (d,e) \cap (b,p)$. In both cases $f \in \langle d,e\rangle \cap (b,c)$.

Suppose next that $d \in (b,c)$. We may assume that $p \neq d$ and, without loss of generality, that $d \in (b,p)$. If $a \in (e,p)$, then there exists a point $f \in (d,e) \cap (a,b)$. If $p \in (a,e)$, then $d \in (e,f)$ for some $f \in (a,b)$. In both cases $f \in \langle d,e\rangle \cap (a,b)$.

The case $d \in (a,c)$ is reduced to the case $d \in (a,b)$ by interchanging b and c. □

Although Proposition 10 may seem rather special, it has some important general consequences. The next lemma is proved by Lenz (1992) under stronger hypotheses.

LEMMA 11 *Let H be a hyperplane. If x_1,x_2,x_3,x_4 are points of $X \setminus H$ such that*

$$(x_1,x_2) \cap H \neq \varnothing, \quad (x_2,x_3) \cap H \neq \varnothing, \quad (x_3,x_4) \cap H \neq \varnothing,$$

then also $(x_4,x_1) \cap H \neq \varnothing$.

Proof Obviously we can assume $x_3 \neq x_1$ and $x_4 \neq x_2$. The lines $<x_1,x_2>$, $<x_2,x_3>$ and $<x_3,x_4>$ intersect H in unique points h_1,h_2 and h_3. Moreover $h_1 \in (x_1,x_2)$, $h_2 \in (x_2,x_3)$ and $h_3 \in (x_3,x_4)$.

Suppose first that x_1,x_2,x_3 are collinear. Then $h_1 = h_2$ and either $x_1 \in (h_2,x_3)$ or $x_3 \in (h_2,x_1)$. If $x_1 \in (h_2,x_3)$ then, by **(P)**, there exists a point $h \in [h_2,h_3] \cap (x_1,x_4)$. If $x_3 \in (h_2,x_1)$ then, by **(C)**, there exists a point $h \in [x_1,x_4]$ such that $h_3 \in [h_2,h]$. If $h_3 = h_2$, then $h_2 \in (x_1,x_4)$. If $h_3 \neq h_2$, then $h \in H \cap (x_1,x_4)$.

Thus we may assume that x_1,x_2,x_3 are not collinear, and hence $h_1 \neq h_2$. Suppose now that x_1,x_3,x_4 are collinear, so that $h_3 \in <x_1,x_3>$. Since $<h_1,h_2,h_3> \neq <x_1,x_2,x_3>$, we must have $h_3 \in <h_1,h_2>$. By Proposition 10, $h_3 \notin (x_1,x_3)$. Since x_1,x_3,x_4 are collinear, it follows that $h_3 \in (x_1,x_4)$.

Thus we may assume that x_1,x_3,x_4 are not collinear and indeed, by similar arguments, we may assume that no three of the points x_1,x_2,x_3,x_4 are collinear.

Suppose next that $<x_2,x_4> \cap H \neq \emptyset$. Then the line $<x_2,x_4>$ intersects H in a unique point h. We show first that $h \notin (x_2,x_4)$. Assume on the contrary that $h \in (x_2,x_4)$. Since x_2,x_3,x_4 are not collinear, it follows from Proposition 10 that $h \notin <h_2,h_3>$. Hence $<h_2,h_3,h> = <x_2,x_3,x_4>$, which is a contradiction.

Thus either $x_4 \in (x_2,h)$ or $x_2 \in (x_4,h)$. If $x_4 \in (x_2,h)$ then, by **(P)**′, there exists a point $h' \in (h,h_1) \cap (x_1,x_4)$. If $x_2 \in (x_4,h)$ then, by Proposition II.2′, there exists a point $h' \in (x_1,x_4)$ such that $h_1 \in (h,h')$.

Thus we may assume that $<x_2,x_4> \cap H = \emptyset$, and similarly also that $<x_1,x_3> \cap H = \emptyset$. By **(P)**′, there exist a point $x \in (x_2,h_3) \cap (x_4,h_2)$ and a point $h \in (x,x_3) \cap (h_2,h_3)$. Since $<x_2,x> \cap H \neq \emptyset$, it follows from the previous part of the proof (with x_4 replaced by x) that $(x_1,x) \cap H \neq \emptyset$. Since $<x,x_4> \cap H \neq \emptyset$, it further follows from the previous part of the proof (with x_2 replaced by x) that $(x_1,x_4) \cap H \neq \emptyset$. □

PROPOSITION 12 *Let H be a hyperplane such that $X \setminus H$ is not convex. Then there exist unique nonempty convex sets H_+ and H_- such that $X \setminus H = H_+ \cup H_-$. Furthermore,*

(i) $H_+ \cap H_- = \emptyset$,

(ii) *if $y \in H_+$ and $z \in H_-$, then $(y,z) \cap H \neq \emptyset$,*

(iii) *$H \cup H_+$ and $H \cup H_-$ are also convex.*

Proof Since $X\setminus H$ is not convex, there exist points $a,b \in X\setminus H$ and $c \in H$ such that $c \in (a,b)$. Let H_+ denote the set of all points $x \in X$ such that $(b,x) \cap H \neq \emptyset$, and let H_- denote the set of all points $x \in X$ such that $(a,x) \cap H \neq \emptyset$. Then $a \in H_+$, $b \in H_-$ and $H_+ \cap H = \emptyset$, $H_- \cap H = \emptyset$.

We are going to show that also $H_+ \cap H_- = \emptyset$. Assume on the contrary that there exists a point $x \in H_+ \cap H_-$. If $x \in \langle a,b\rangle$ then $c \in (a,x) \cap (b,x)$, since $\langle a,b\rangle \cap H = \{c\}$. But this is impossible, since also $c \in (a,b)$. Thus a,b,x are not collinear. If (a,x), (b,x) intersect H in h,h' respectively, then $c \notin \langle h,h'\rangle$ by Proposition 10. Hence $\langle c,h,h'\rangle = \langle a,b,x\rangle$, which is a contradiction.

We show next that H_- is convex. Assume on the contrary that there exist points $x',x'' \in H_-$ and a point $x \in (x',x'')$ such that $x \notin H_-$. Then there exist points $h',h'' \in H$ such that $h' \in (a,x')$, $h'' \in (a,x'')$. Hence a,x',x'' are not collinear and $h' \neq h''$. By Proposition II.17′ the segment (h',h'') contains a point $h \in (a,x)$. Then $h \in H$ and $x \in H_-$, which is a contradiction. Thus H_- is convex, and similarly also H_+.

It follows at once from Lemma 11 that if $y \in H_+$ and $z \in H_-$, then $(y,z) \cap H \neq \emptyset$. It will now be shown that $X = H \cup H_+ \cup H_-$. Since H is a hyperplane, it is sufficient to show that $X' := H \cup H_+ \cup H_-$ is affine. In fact it is enough to show that if $x' \in H_+$, $x'' \in X'$ and $x \in \langle x',x''\rangle$, then $x \in X'$. Since $x' \in H_+$, the segment (b,x') contains a point $h' \in H$.

Suppose first that $x'' = h'' \in H$. If $h'' \in (x,x')$ then $x \in H_-$, since $x' \in H_+$. If $x \in (x',h'')$ then, by **(P)**, there exists a point $h \in [h',h''] \cap (b,x)$ and hence $x \in H_+$. If $x' \in (x,h'')$ and b,x,x' are collinear, then $h' = h'' \in (b,x)$ and hence $x \in H_+$. If $x' \in (x,h'')$ and b,x,x' are not collinear then, by Proposition II.2′, there exists a point $h \in (b,x)$ such that $h' \in (h,h'')$ and again $x \in H_+$. This proves that a line containing a point of H and a point of H_+ is entirely contained in X', and similarly a line containing a point of H and a point of H_- is entirely contained in X'.

Suppose next that $x'' \in H_+ \cup H_-$ and the line $\langle x',x''\rangle$ contains no point of H. Then $x'' \in H_+$, since $x'' \in H_-$ would imply $(x',x'') \cap H \neq \emptyset$. Hence the segment (b,x'') contains a point $h'' \in H$. If $x \in (x',x'')$ then $x \in H_+$, since H_+ is convex. Now consider the case $x' \in (x,x'')$. By **(P)′**, there exists a point $y \in (b,x') \cap (h'',x)$. Moreover $y \notin H$. If $h' \in (b,y)$, then the segment (b,x) contains a point h such that $h' \in (h,h'')$ and hence $x \in H_+$. If $h' \in (x',y)$, then the segment (x,x') contains a point h such that $h' \in (h,h'')$ and hence $x \in X'$ by what we have already proved. Since the same argument applies in the case $x'' \in (x,x')$, this completes the proof that $X = H \cup H_+ \cup H_-$.

It now follows that $H \cup H_+$ and $H \cup H_-$ are convex. For if (say) $x_1 \in H$ and $x_2 \in H_+$, then $(x_1,x_2) \cap H = \emptyset$ and hence $(x_1,x_2) \cap H_- = \emptyset$.

Finally, suppose $X \setminus H$ is the union of two nonempty convex sets G_+ and G_-. We may assume the notation chosen so that $G_+ \cap H_+ \neq \emptyset$. But then $G_+ \subseteq H_+$, by (ii). Hence $G_- \cap H_- \neq \emptyset$ and so, in the same way, $G_- \subseteq H_-$. Since $G_+ \cup G_- = H_+ \cup H_-$, we must actually have $G_+ = H_+$ and $G_- = H_-$. □

Proposition 12 says that a hyperplane has two 'sides', if its complement is not convex. The convex sets H_+ and H_- in the statement of Proposition 12 will be called the *open half-spaces* associated with the hyperplane H, and the convex sets $H_+ \cup H$ and $H_- \cup H$ will be called the *closed half-spaces* associated with H. (When $X \setminus H$ is convex, we may call $X \setminus H$ the open half-space and X the closed half-space associated with the hyperplane H. This situation arises if, for example, X is itself the closed half-space associated with a hyperplane H of a real vector space.) Our topological terminology will be given more justification in Chapter V.

Lemma 9 can be generalized to higher dimensions in the following way:

PROPOSITION 13 *Let H_1,H_2,H_3 be three distinct hyperplanes which themselves have a common hyperplane Z. If $x_k \in H_k \setminus Z$ $(k = 1,2,3)$, and if*

$$[x_2,x_3] \cap H_1 = \emptyset, \quad [x_1,x_3] \cap H_2 = \emptyset,$$

then $[x_1,x_2] \cap H_3 \neq \emptyset$.

Proof Assume on the contrary that $[x_1,x_2] \cap H_3 = \emptyset$. Then x_1,x_2,x_3 are not collinear and hence $Z \neq \emptyset$. Moreover, by Lemma 9,

$$\langle x_1,x_2,x_3\rangle \cap Z = \emptyset.$$

Hence, by Proposition 8, for any $z \in Z$ we have

$$(*) \qquad x_1 \notin \langle x_2,x_3,z\rangle, \; x_2 \notin \langle x_1,x_3,z\rangle, \; x_3 \notin \langle x_1,x_2,z\rangle.$$

Since $x_3 \in X = \langle x_1,x_2,Z\rangle$, it follows from Proposition 7 that $x_3 \in \langle y_1,y_2\rangle$, where $y_j \in [x_1,x_2,z_j]$ for some $z_j \in Z$ $(j = 1,2)$. Moreover (*) implies that $z_1 \neq z_2$, $y_j \notin [x_1,x_2]$ for $j = 1,2$, and $x_3 \notin [y_1,y_2]$. Without loss of generality we may suppose $y_1 \in (y_2,x_3)$.

Put

$$X' = \langle x_1,x_2,z_1,z_2\rangle .$$

Then dim $X' = 3$ and $x_3 \in X'$. Furthermore $H_k' = \langle x_k,z_1,z_2\rangle$ ($k = 1,2,3$) are distinct planes in X' with the common line $Z' = \langle z_1,z_2\rangle$. Moreover

$$[x_2,x_3] \cap H_1' = \emptyset, \quad [x_1,x_3] \cap H_2' = \emptyset, \quad [x_1,x_2] \cap H_3' = \emptyset.$$

Since $y_2 \in [x,z_2]$ for some $x \in [x_1,x_2]$ and $y_2 \notin [x_1,x_2]$, we must have $[y_2,z_2] \cap \langle x_1,x_2\rangle = \emptyset$. Since

$$\langle x_1,x_2,z_1\rangle \cap \langle x_1,x_2,z_2\rangle = \langle x_1,x_2\rangle,$$

it follows that $[y_2,z_2] \cap \langle x_1,x_2,z_1\rangle = \emptyset$. Thus y_2 and z_2 lie in the same open half-space of X' associated with the plane $\langle x_1,x_2,z_1\rangle$. Since

$$y_1 \in (y_2,x_3) \cap \langle x_1,x_2,z_1\rangle,$$

it follows that x_3 and z_2 lie in different open half-spaces of X' associated with the plane $\langle x_1,x_2,z_1\rangle$. Thus there exists a point

$$u \in (x_3,z_2) \cap \langle x_1,x_2,z_1\rangle.$$

Since $[x_2,x_3] \cap H_1' = \emptyset$, it follows that

$$[x_2,u] \cap \langle x_1,z_1\rangle = \emptyset.$$

Similarly from $[x_1,x_3] \cap H_2' = \emptyset$ we obtain

$$[x_1,u] \cap \langle x_2,z_1\rangle = \emptyset,$$

and $[x_1,x_2] \cap H_3' = \emptyset$ immediately implies

$$[x_1,x_2] \cap \langle u,z_1\rangle = \emptyset.$$

Consequently, by Lemma 9, $z_1 \notin \langle x_1,x_2,u\rangle$. But $u \notin \langle x_1,x_2\rangle$, since $z_2 \notin \langle x_1,x_2,x_3\rangle$ and hence, by Proposition 8, $z_1 \in \langle x_1,x_2,u\rangle$. Thus we have a contradiction. □

The *join* of two convex geometries was defined in Chapter II and, in the terminology of this chapter, it was shown there that the join of two linear geometries is again a linear geometry. A linear geometry will be said to be *irreducible* if it is not itself the join of two linear geometries. Since irreducible linear geometries are the building blocks of which all linear geometries are composed, it is of interest to determine their nature. To this task we now turn.

LEMMA 14 *Let x,y_1,y_2 be non-collinear points such that $\langle x,y_1\rangle \neq \{x,y_1\}$ and $\langle x,y_2\rangle \neq \{x,y_2\}$. If $[y_1,y_2] = \{y_1,y_2\}$, then there exist points z_1,z_2 such that $x \in (y_1,z_1) \cap (y_2,z_2)$ and the plane $\langle x,y_1,y_2\rangle$ contains only the five points x,y_1,y_2,z_1,z_2.*

Proof Let $z_k \in \langle x,y_k\rangle$ with $z_k \neq x,y_k$ $(k = 1,2)$. Assume first that $z_2 \in (x,y_2)$. If $y_1 \in (x,z_1)$ then, by **(P)′**, there exists a point $w \in (y_1,y_2) \cap (z_1,z_2)$. If $x \in (y_1,z_1)$ then, by Proposition II.2′, $z_2 \in (y,z_1)$ for some $y \in (y_1,y_2)$. If $z_1 \in (x,y_1)$ then, by **(P)′**, there exists a point $u \in (y_1,z_2) \cap (y_2,z_1)$ and, by Proposition II.2′, $u \in (x,v)$ for some $v \in (y_1,y_2)$. Since $(y_1,y_2) = \emptyset$ we conclude that $(x,y_2) = \emptyset$, and likewise $(x,y_1) = \emptyset$.

Assume next that $y_2 \in (x,z_2)$. If $x \in (y_1,z_1)$, then $y_2 \in (z_1,u)$ for some $u \in (y_1,z_2)$ and there exists a point $v \in (x,u) \cap (y_1,y_2)$. If $y_1 \in (x,z_1)$, then there exist a point $u' \in (y_1,z_2) \cap (y_2,z_1)$ and a point $v' \in (x,u) \cap (y_1,y_2)$. Since $(y_1,y_2) = \emptyset$ we conclude that $x \in (y_2,z_2)$, and likewise $x \in (y_1,z_1)$.

Assume now that the line $\langle x,y_1\rangle$ contains a point $z \neq x,y_1,z_1$. Then in the same way $x \in (y_1,z)$ and by interchanging z and z_1, if necessary, we may suppose that $z \in (x,z_1)$. But then $z \in (y_2,z')$ for some $z' \in (z_1,z_2)$ and hence, by Proposition II.19′, $x \in (y',z')$ for some $y' \in (y_1,y_2)$. We conclude that $\langle x,y_1\rangle = \{x,y_1,z_1\}$, and similarly $\langle x,y_2\rangle = \{x,y_2,z_2\}$.

Assume next that the line $\langle y_1,y_2\rangle$ contains a point $z \neq y_1,y_2$. Without loss of generality we may suppose $y_2 \in (y_1,z)$. Then $x \in (z,u)$ for some $u \in (y_1,z_2)$ and there exists a point $v \in (u,y_2) \cap (x,y_1)$. Since $(x,y_1) = \emptyset$, we conclude that $\langle y_1,y_2\rangle = \{y_1,y_2\}$.

Assume finally that the plane $\langle x,y_1,y_2\rangle$ contains a point $z \neq x,y_1,y_2,z_1,z_2$. Then

$$[y_1,y_2] \cap \langle x,z\rangle = \emptyset,$$

since $(y_1,y_2) = \emptyset$ and $z \notin \langle x,y_k\rangle$ $(k = 1,2)$. Furthermore

$$[x,y_1] \cap \langle y_2,z\rangle = \emptyset,$$

since $(x,y_1) = \emptyset$, $z \notin \langle y_1,y_2\rangle$ and $z \notin \langle x,y_2\rangle$. Similarly

$$[x,y_2] \cap \langle y_1,z\rangle = \emptyset.$$

But this contradicts Lemma 9. □

COROLLARY 15 *If not all points of the linear geometry X are collinear and if every line contains at least three points then, for any distinct points $a,b \in X$, $(a,b) \neq \varnothing$.* □

Irreducible linear geometries are completely characterized by the following proposition:

PROPOSITION 16 *If X is an irreducible linear geometry, then exactly one of the following alternatives holds:*

(i) *X contains only one point,*

(ii) *X contains at least three points and all points of X are collinear,*

(iii) *not all points of X are collinear and $[a,b] \neq \{a,b\}$ for any distinct points $a,b \in X$,*

(iv) *not all points of X are collinear and there exists $z \in X$ such that every line which does not contain z has only two points, and every line which contains z has exactly two other points with z in the segment determined by them.*

Proof It is clear that X is irreducible in each of the cases (i)–(iv), and that no two of these cases can coexist. To prove the proposition we will assume that none of (i)–(iv) holds and derive a contradiction.

Since X is irreducible, our assumptions imply that X contains at least three points and not all points are collinear. If A is a nonempty proper affine subset of X then, for some $a \in A$ and some $x \in X \setminus A$, the line $\langle a,x \rangle$ contains more than two points (since otherwise $X \setminus A$ would be affine and X would be the join of A and $X \setminus A$).

In particular some line in X contains at least three points. It follows from Hausdorff's maximality theorem that there is an affine subset A of X, every line of which contains at least three points, but which is not properly contained in another affine subset with the same property. Since (iii) does not hold, it follows from Corollary 15 that $A \neq X$. Hence there exist points $x \in X \setminus A$ and $z \in A$ such that $\langle z,x \rangle$ contains at least three points.

Assume there exist distinct points $x_1,x_2 \in \langle A \cup x \rangle \setminus A$ such that

$\langle x_1,x_2\rangle = \{x_1,x_2\}$. Then $\langle A \cup x\rangle = \langle A \cup x_1\rangle$ and it follows from Proposition 7 that $x_2 \in \langle x',x''\rangle$, where $x' \in [a',x_1]$ and $x'' \in [a'',x_1]$ for some $a',a'' \in A$. Since $x_1 \notin \langle a',a''\rangle$ and $x_2 \notin \langle a',a''\rangle \cup \langle a',x_1\rangle \cup \langle a'',x_1\rangle$, we may suppose that $x' \in (a',x_1)$ and $x'' \in [a'',x_1)$. It follows from Lemma 14 that $x' \in (x_2,x'')$ and that the plane $\langle a',a'',x_1\rangle$ contains only five points. Hence $x'' = a''$ and the line $\langle a',a''\rangle$ contains only two points, which is a contradiction.

We conclude that if x_1,x_2 are distinct points of $\langle A \cup x\rangle \setminus A$, then $\langle x_1,x_2\rangle \neq \{x_1,x_2\}$. But, since A is maximal, there exist distinct $y_1,y_2 \in \langle A \cup x\rangle$ such that $\langle y_1,y_2\rangle = \{y_1,y_2\}$. We may choose the notation so that $y_1 = a \in A$ and $y_2 \notin A$. If $\langle a,x\rangle \neq \{a,x\}$, then $y_2 \neq x$ and $\langle y_2,x\rangle \neq \{y_2,x\}$. Hence, by Lemma 14, there exists a point y such that $x \in (a,y)$ and $\langle y,y_2\rangle = \{y,y_2\}$. Since $y \notin A$, this is a contradiction. Thus we may take $y_2 = x$.

Since the lines $\langle z,x\rangle$ and $\langle z,a\rangle$ contain at least three points, it now follows from Lemma 14 that there exist points a',x' such that $z \in (a,a') \cap (x,x')$ and the plane $\langle z,a,x\rangle$ contains only the points z,a,x,a',x'. Hence, by Corollary 15, $A = \{z,a,a'\}$.

Choose any point $b \notin \langle z,a\rangle$. If $\langle z,b\rangle \neq \{z,b\}$, then any affine subset properly containing $\langle z,b\rangle$ has a line containing only two points. In particular the plane $\langle z,a,b\rangle$ has a line containing only two points. The preceding argument, with A replaced by $\langle z,a\rangle$ and x by b, shows that there exists a point b' such that $z \in (b,b')$ and $\langle z,a,b\rangle = \{z,a,b,a',b'\}$.

Let C be the union of z with the set of all points $c \in X$ such that $\langle z,c\rangle \neq \{z,c\}$. It follows from what has been said that C is affine. Moreover $C \neq X$, since (iv) does not hold. Hence there exist points $p \in X \setminus C$ and $q \in C$ such that $\langle p,q\rangle \neq \{p,q\}$. Since $\langle z,p\rangle = \{z,p\}$, we must have $q \neq z$ and $\langle z,q\rangle \neq \{z,q\}$. Hence $\langle z,q\rangle = \{z,q,q'\}$, where $z \in (q,q')$. But since $\langle p,q\rangle \neq \{p,q\}$, it follows from Lemma 14 also that $q \in (z,q')$. Thus we have a contradiction. □

An irreducible linear geometry for which case (iv) of Proposition 16 holds may be called a *multi-cross* with (uniquely determined) *centre* z, since each plane containing z has exactly four other points, u_1,u_2,v_1,v_2 say, with $z \in (u_1,v_1) \cap (u_2,v_2)$. If V is a vector space over an ordered division ring D with basis $\{e_i\}_{i\in I}$, then the restriction of V to the set $X = \{O,e_i,-e_i\}_{i\in I}$ is a multi-cross with centre O.

3 HELLY'S THEOREM AND ITS RELATIVES

It will now be shown that in a linear geometry Proposition II.13 admits a stronger formulation:

PROPOSITION 17 *In a linear geometry the classes of Helly sets, Radon sets and affine independent sets all coincide.*

Proof By Proposition II.13 we need only show that every affine independent set is a Radon set, and that every Helly set is affine independent. Moreover we may restrict attention to finite sets.

Assume first that $S = \{s_1,...,s_n\}$ is an affine independent set, but not a Radon set. Then we may choose the notation so that $[S_1] \cap [S_2] \neq \emptyset$, where $S_1 = \{s_1,...,s_m\}$ and $S_2 = \{s_{m+1},...,s_n\}$ for some m with $1 \leq m < n$. Let $x \in [S_1] \cap [S_2]$. Then $x \neq s_1$, since $[S_2] \subseteq \langle S \setminus s_1 \rangle$ and S is affine independent. On the other hand $x \in \langle s_1,...,s_m \rangle$. Hence we can choose k, with $1 \leq k < m$, so that $x \notin \langle s_1,...,s_k \rangle$ but $x \in \langle s_1,...,s_{k+1} \rangle$. Then $s_{k+1} \in \langle x,s_1,...,s_k \rangle$, by Proposition 8. Since $x \in \langle s_{m+1},...,s_n \rangle$, it follows that $s_{k+1} \in \langle S \setminus s_{k+1} \rangle$, which is a contradiction.

To show that a finite Helly set is affine independent we use induction on the cardinality of the set. Since the result is true for singletons, it is sufficient to establish the following assertion:

(#) If $T = \{x_1,...,x_n\}$ is an affine independent set and if $x_{n+1} \in \langle T \rangle \setminus T$, then $S = T \cup x_{n+1}$ is not a Helly set.

This is certainly true for $n = 2$, since either $x_3 \in [x_1,x_2]$ or $x_1 \in [x_2,x_3]$ or $x_2 \in [x_1,x_3]$. We assume that $n > 2$ and the assertion holds for all smaller values of n. We may assume also that $x_{n+1} \notin [T]$, since $x_{n+1} \in [T]$ implies $x_{n+1} \in [S \setminus x_i]$ for $i = 1,...,n+1$.

By Proposition 7 we have $x_{n+1} \in \langle y,z \rangle$, where $y,z \in [T]$ and $y \neq z$. By Proposition II.14, there exist $x_j,x_k \in T$ such that $y \in [z \cup (T \setminus x_j)]$ and $z \in [y \cup (T \setminus x_k)]$. Consequently $y \in [z,y']$, where $y' \in [T \setminus x_j]$, and $z \in [y,z']$, where $z' \in [T \setminus x_k]$. Then $y,z \in [y',z']$, by **(L2)**, and hence $x_{n+1} \in \langle y',z' \rangle$.

Since $x_{n+1} \notin [T]$, we have $x_{n+1} \notin [y',z']$. Without loss of generality, suppose $z' \in (y',x_{n+1})$. We may assume $z' \notin [T \setminus x_j]$ since otherwise, by the induction hypothesis, $S \setminus x_j$ (and hence also S) is not a Helly set. Then $j \neq k$ and $y' \in [x_k,u]$, where $u \in [T\setminus(x_j \cup x_k)]$. Hence $z' \in [x_k,x_{n+1},u]$, and so $z' \in [u,v]$, where $v \in [x_k,x_{n+1}]$. Since $u \in [T\setminus(x_j \cup x_k)]$ and $z' \in [T\setminus x_k]$, it follows that

$$v \in \langle z',u\rangle \subseteq \langle T\setminus x_k\rangle.$$

If $v \in T\setminus x_k$, say $v = x_i$, then $x_{n+1} \in \langle x_i,x_k\rangle$ and we are reduced to the case $n = 2$. If $v \notin T\setminus x_k$ then, by the induction hypothesis, the set

$$R := v \cup (T\setminus x_k)$$

is not a Helly set. Thus there exists a point $x \in [T\setminus x_k]$ such that $x \in [R\setminus x_j]$ for all $j \in \{1,\ldots,n\}$ with $j \neq k$. Since $v \in [x_k,x_{n+1}]$, it follows that $x \in [S\setminus x_i]$ for $i = 1,\ldots,n+1$. □

COROLLARY 18 *For any set $S \subseteq X$ and any $x \in [S]$, there exists a finite affine independent set $F \subseteq S$ such that $x \in [F]$.*

Proof Let F be a finite subset of S such that $x \in [F]$, but no proper subset of F has the same property. Then F is a Carathéodory set. Hence F is a Helly set, by Proposition II.15, and thus also an affine independent set, by Proposition 17. □

Proposition 17 no longer holds if all the axioms for a linear geometry are satisfied with the exception of **(P)**. This is shown by the example illustrating the independence of the axiom **(P)** in Example II.10. In this example $S = \{a,a',b,b'\}$ is a Helly set, but it is not affine independent since $a' \in \langle a,c\rangle \subseteq \langle a,b,b'\rangle$.

If S is an affine independent subset of $\mathbb{R}^d$, then $|S| \leq d+1$. Consequently, by Proposition 17, any Helly set in $\mathbb{R}^d$ has at most $d+1$ elements and hence, by Proposition I.10, a finite family of at least $d+1$ convex sets in $\mathbb{R}^d$ has nonempty intersection if every subfamily of $d+1$ sets has nonempty intersection. This was first proved by Helly (1923).

By Proposition 17 also, any Radon set in $\mathbb{R}^d$ has at most $d+1$ elements. Hence any subset S of $\mathbb{R}^d$ with $|S| > d+1$ has disjoint subsets S_1,S_2 such that $[S_1] \cap [S_2] \neq \emptyset$. This was first proved by Radon (1921), who used it to give the first published proof of Helly's theorem.

Similarly it follows from Corollary 18 that if $S \subseteq \mathbb{R}^d$ and $x \in [S]$, then there exists a subset F of S with $|F| \leq d+1$ such that $x \in [F]$. This was first proved by Carathéodory (1911).

The formulation of Proposition 17 has several advantages over the classical formulations: the condition given is both necessary and sufficient, the result holds in vector spaces over any ordered division ring, and the vector space need not be finite-dimensional. (For Corollary 18, see also Proposition V.2 below.)

There is an extensive literature dealing with generalizations, modifications and analogues of the theorems of Helly, Radon and Carathéodory; see Danzer *et al.* (1963), from which the present section takes its title, and Eckhoff (1979,1993). We restrict ourselves to describing some interesting recent results in this area and drawing attention to some open problems.

Bárány (1982) has proved the following theorem, which contains Carathéodory's theorem as a special case ($S_1 = ... = S_{d+1} = S$). However, the proof uses Carathéodory's theorem itself.

BÁRÁNY'S THEOREM *If a point x belongs to the convex hulls of d+1 sets $S_1,...,S_{d+1}$ in $\mathbb{R}^d$, then there exist points $x_i \in S_i$ $(i = 1,...,d+1)$ such that x belongs to the convex hull of $\{x_1,...,x_{d+1}\}$.*

Proof We may assume, without loss of generality, that $x = 0$ and that $\mathbb{R}^d$ is equipped with the Euclidean norm. Furthermore, by Carathéodory's theorem, we may suppose that each set S_i contains at most d+1 points. Then the collection $\mathcal{S}$ of all sets $S = \{x_1,...,x_{d+1}\}$ with $x_i \in S_i$ $(i = 1,...,d+1)$ is finite. For any $S \in \mathcal{S}$, put

$$\mathrm{d}(S) = \inf\{\|x\| : x \in [S]\}.$$

Since $[S]$ is compact, we have $\mathrm{d}(S) = \|z\|$ for some $z \in [S]$. We wish to show that $\mathrm{d}(S) = 0$ for some $S \in \mathcal{S}$. Since $\mathcal{S}$ is finite, it is sufficient to show that if $\mathrm{d}(S) > 0$, then there exists an $S' \in \mathcal{S}$ such that $\mathrm{d}(S') < \mathrm{d}(S)$.

If S is not affine independent then, by Carathéodory's theorem, z is a convex combination of at most d points of S. If S is affine independent, then the nearest point of $[S]$ to the origin lies on some proper face. In either case $z \in [S \setminus x_j]$ for some $j \in \{1,...,d+1\}$.

The origin cannot be separated from S_j by any hyperplane, since $0 \in [S_j]$. Consequently the open half-space $\{x \in \mathbb{R}^d: (x - z,z) < 0\}$, which contains the origin, must also contain a point $x_j' \in S_j$. Let S' be the set which is obtained from S by replacing x_j by x_j' and retaining the other x_i. Then $[S']$ contains the segment $[x_j',z]$, since it contains both x_j' and z. But, for small $\lambda > 0$,

$$\|\lambda x_j' + (1-\lambda)z\|^2 = \|z\|^2 + 2\lambda(x_j' - z, z) + O(\lambda^2)$$
$$< \|z\|^2.$$

Hence $d(S') < d(S)$. □

Another proof of Bárány's theorem is given by Kovijanić (1994). It may be asked if Bárány's theorem admits the following extension to linear geometries:

Let $S_1,\ldots,S_m$ be subsets of a linear geometry X such that $\dim \langle S_1,\ldots,S_m\rangle < m$. *If $x \in [S_1] \cap \ldots \cap [S_m]$, does there exist an affine independent set $\{x_1,\ldots,x_n\}$, with $x_j \in S_{i_j}$ for distinct integers $i_1,\ldots,i_n \in \{1,\ldots,m\}$, such that $x \in [x_1,\ldots,x_n]$?*

Tverberg (1966) established a strong generalization of Radon's theorem, and a simpler proof has been given by Tverberg and Vrecica (1993). The following deduction of Tverberg's theorem from Bárány's theorem, due to Onn, is contained in Sarkaria (1992).

TVERBERG'S THEOREM *If a set $S \subseteq \mathbb{R}^d$ contains $(r-1)(d+1) + 1$ distinct points, for some integer $r > 1$, then S can be expressed as the union of r pairwise disjoint subsets whose convex hulls have a common point.*

Proof We can identify $\mathbb{R}^d$ with the hyperplane $\xi_1 + \ldots + \xi_{d+1} = 1$ in $\mathbb{R}^{d+1}$. Let $v_0,v_1,\ldots,v_n$, where $n = (r-1)(d+1)$, be the points of S in this identification. With each point $v_i \in S$ we associate a set $S_i' = \{M_{i1},\ldots,M_{ir}\}$ of $(d+1)\times(r-1)$ matrices defined in the following way: if $1 \leq k < r$ the matrix M_{ik} has v_i as its k-th column and zeros elsewhere, whereas every column of M_{ir} is $-v_i$. If we also identify the set of all $(d+1)\times(r-1)$ matrices with $\mathbb{R}^n$, we have $n+1$ sets $S_i' \subseteq \mathbb{R}^n$ with $|S_i'| = r$. Moreover, the origin is contained in the convex hull of every set S_i', since $M_{i1} + \ldots + M_{ir} = 0$. Consequently, by Bárány's theorem, there exist a matrix $M_{ik_i} \in S_i'$ and $\lambda_i \geq 0$ with $\sum_{i=0}^n \lambda_i = 1$ such that

$$\sum_{i=0}^n \lambda_i M_{ik_i} = 0.$$

Let I_j denote the set of all $i \in \{0,1,\ldots,n\}$ for which $k_i = j$. Then we can rewrite the last equation in the form

$$\sum_{j=1}^r \sum_{i\in I_j} \lambda_i M_{ij} = 0.$$

Equivalently,

$$\sum_{i \in I_1} \lambda_i v_i = \sum_{i \in I_2} \lambda_i v_i = \ldots = \sum_{i \in I_r} \lambda_i v_i.$$

Put $\mu_j = \sum_{i \in I_j} \lambda_i$. Since each v_i lies on the hyperplane $\xi_1 + \ldots + \xi_{d+1} = 1$ and $\sum_{i=0}^{n} \lambda_i = 1$, it follows that $\mu_1 = \mu_2 = \ldots = \mu_r = 1/r$. Hence, if we denote by S_j the set of all v_i with $i \in I_j$, the given set S has a partition into pairwise disjoint subsets $S_1,\ldots,S_r$ whose convex hulls $[S_1],\ldots,[S_r]$ have a common point. □

Tverberg's theorem is best possible, in the sense that the number $(r-1)(d+1) + 1$ in its statement cannot be replaced by any smaller number. To understand the significance of this number, define the r-th *Tverberg number* τ_r of $\mathbb{R}^d$ to be the maximum cardinality of any set in $\mathbb{R}^d$ such that, for every partition of the set into r pairwise disjoint subsets, the intersection of their convex hulls is always empty. In particular, τ_2 is the maximum cardinality of any Radon set in $\mathbb{R}^d$. Then Tverberg's theorem can be simply restated in the form: $\tau_r = (r-1)\tau_2$.

In the definition of Tverberg numbers, the space $\mathbb{R}^d$ can be replaced by an arbitrary alignment. Eckhoff (1979) has conjectured that the inequality $\tau_r \leq (r-1)\tau_2$ holds in any alignment for which τ_2 is finite. This sweeping generalization of Tverberg's theorem (first formulated by Calder (1971) for Example 4 of Chapter II) has been neither proved nor disproved, although it is known that $\tau_2 < \infty$ implies $\tau_r < \infty$ for every $r > 2$. In the context of the present chapter it may be asked: *does Eckhoff's conjecture hold at least in any linear geometry?*

There is another conjectured generalization of Tverberg's theorem within $\mathbb{R}^d$ itself: if a set $S \subseteq \mathbb{R}^d$ contains $r(d+1)$ distinct points, partitioned into $d+1$ subsets $A_1,\ldots,A_{d+1}$ containing r points each, can S be expressed as the union of r pairwise disjoint subsets $S_1,\ldots,S_r$ whose convex hulls have a common point in such a way that every subset S_i $(i = 1,\ldots,r)$ contains exactly one point from each subset A_k $(k = 1,\ldots,d+1)$? A proof for $r = 2$ is contained in Bárány and Larman (1992). They also prove the conjecture for $d = 2$.

IV
Linearity (continued)

We show here that faces may be given yet another characterization in a linear geometry, and that *polytopes* possess most of their familiar properties in $\mathbb{R}^d$. A notable omission is the Euler–Poincaré relation. It is shown also that from a given linear geometry further *factor geometries* may be derived.

1 FACES

Throughout this chapter we assume that a set X is given on which a linear geometry is defined. The notion of 'face' of a convex set was defined in any aligned space in Chapter I, and in Chapter II a simpler characterization was given in any convex geometry. Yet another characterization will now be given in any linear geometry:

PROPOSITION 1 *A set A is a face of a convex set C if and only if $C \cap \langle A \rangle = A$ and $C \setminus A$ is convex.*

Proof Let A be a set such that $C \cap \langle A \rangle = A$ and $C \setminus A$ is convex. Then $A \subseteq C$ and A is convex, since it is the intersection of two convex sets. If $a \in (c,c')$, where $a \in A$ and $c,c' \in C$, then at least one of c,c' is in A, since $C \setminus A$ is convex, and in fact they both are, since $C \cap \langle A \rangle = A$. Thus A is a face of C, by Proposition II.6.

On the other hand, if A is a face of C then $C \setminus A$ is convex, by Proposition I.5. Put $B = C \cap \langle A \rangle$. Then A is a face of B, by Proposition I.7, since B is convex and $A \subseteq B \subseteq C$. By Proposition III.7, if $b \in B$ then $b \in \langle a,a' \rangle$ for some $a,a' \in A$. If $b \in [a,a']$ then $b \in A$, since A is convex. If $a' \in (a,b)$ or $a \in (a',b)$ then $b \in A$, since A is a face of B. Thus $B = A$. □

COROLLARY 2 *Let A and B be faces of a convex set C. Then A properly contains B if and only if $\langle A\rangle$ properly contains $\langle B\rangle$.* □

PROPOSITION 3 *If a convex set C is contained in a closed half-space of $\langle C\rangle$ associated with a hyperplane H, then $C \cap H$ is a face of C.*

Proof Put $D = C \cap H$ and let H_+ be the open half-space of $\langle C\rangle$ associated with H such that $C \subseteq H \cup H_+$. Then $C \cap \langle D\rangle = D$, since $D \subseteq C \cap \langle D\rangle \subseteq C \cap H$, and $C \setminus D$ is convex, since $C \setminus D = C \cap H_+$. □

PROPOSITION 4 *Let C be a convex set and H_+ an open half-space of $\langle C\rangle$ associated with a hyperplane H. Also, let A be a subset of $B = C \cap (H \cup H_+)$ such that $A \cap H$ is a face of C (possibly $\emptyset$). Then A is a face of C if and only if A is a face of B.*

Proof By Proposition I.7 we need only show that if A is a face of B, then A is also a face of C. Put $H_- = \langle C\rangle \setminus (H \cup H_+)$.

We show first that $C \cap \langle A\rangle = A$. Since $B \cap \langle A\rangle = A$, it is enough to show that if $y \in C \cap H_-$, then $y \notin \langle A\rangle$. Assume on the contrary that $y \in \langle A\rangle$. Then $y \in \langle x,z\rangle$ for some $x,z \in A$ and we may choose the notation so that $z \in (x,y)$. Hence there exists a point $w \in (y,z] \cap H$. Moreover $w \notin A$, since $A \cap H$ is a face of C. Hence $w \neq z$ and $z \in (x,w)$. But, since A is a face of B, this contradicts $w \notin A$.

We show next that $C \setminus A$ is convex. Assume on the contrary that, for some points $x,y \in C \setminus A$, there exists a point $z \in (x,y) \cap A$. Since $B \setminus A$ is convex, we may assume that $y \in H_-$. Then $x \in H_+$, since $H \cup H_-$ is convex. Hence there exists a point $w \in (x,y) \cap H$, and $z \in [w,x)$, which leads to a contradiction in the same way as before. □

PROPOSITION 5 *Let C be a convex set and S any subset of C. Then the subset A_S of C, consisting of the union of $C \cap \langle S\rangle$ with the set of all points $x,x' \in C$ such that $(x,x') \cap \langle S\rangle \neq \emptyset$, is a face of C. Moreover A_S is contained in every face of C which contains S.*

Proof We show first that A_S is convex. Suppose $z \in (x,y)$, where x,y are distinct elements of A_S with $x \notin \langle S\rangle$ and $z \notin \langle S\rangle$. Then $z \in C$ and there exist points $x' \in A_S$, $a \in \langle S\rangle$ such that $a \in (x,x')$. If a,x,y are collinear, then $a \in (z,z')$ for some $z' \in \{x,y,x'\} \subseteq C$ and so $z \in A_S$. Hence we now suppose that a,x,y are not collinear. Then, by (P)′, there exists a point $c \in (a,y) \cap (x',z)$. If $y \in \langle S\rangle$, then $c \in \langle S\rangle$ and hence $z \in A_S$. If $y \notin \langle S\rangle$, there exist $y' \in A_S$ and $b \in \langle S\rangle$ such that $b \in (y,y')$.

If $b = a$ then, by Proposition II.19′, $a \in (z,z')$ for some $z' \in (x',y')$ and hence $z \in A_S$. Thus we now suppose $b \neq a$ and hence $y \notin \langle a,b\rangle$. By (P)′, there exists a point $d \in (a,b) \cap (c,y')$. Moreover $d \in C \cap \langle S\rangle$, since $a,b \in C \cap \langle S\rangle$. Thus we may suppose $d \neq z$. We may also suppose that $y' \notin \langle x',z\rangle$, since otherwise $d \in (z,x')$ or $d \in (z,y')$. Then, by Proposition II.2′, there exists a point $e \in (x',y')$ such that $d \in (e,z)$. Thus $z \in A_S$ in every case.

We now show that A_S is a face of C. Suppose $x \in A_S$ and $x \in (c,c')$, where $c,c' \in C$. If $x \in \langle S\rangle$ then $c,c' \in A_S$, by the definition of A_S. If $x \notin \langle S\rangle$, then there exist $x' \in A_S$ and $a \in \langle S\rangle$ such that $a \in (x,x')$. If a,c,c' are collinear, then $c = a$ or $a \in (c,x)$ or $a \in (c,x')$. In any event $c \in A_S$, and likewise $c' \in A_S$. Thus we now suppose that a,c,c' are not collinear. Then, by Proposition II.2′, $a \in (c,y)$ for some $y \in (c',x')$. Hence $c \in A_S$, and likewise $c' \in A_S$.

It follows at once from Proposition 1 that any face of C which contains S must also contain A_S. □

COROLLARY 6 *Let C be a convex set and $a \in C$. Then the set A_a consisting of a and all points $x,x' \in C$ such that $a \in (x,x')$ is a face of C. Moreover A_a is contained in every face of C which contains a.* □

A face F of a convex set C will be said to be a *facet* of C if $\langle F\rangle$ is a hyperplane of $\langle C\rangle$, and an *edge* of C if $\dim \langle F\rangle = 1$.

PROPOSITION 7 *Three distinct facets of a convex set cannot themselves have a common nonempty facet.*

Proof Assume on the contrary that three distinct facets F_1,F_2,F_3 of a convex set C have a common nonempty facet F. Then $\langle F_1\rangle,\langle F_2\rangle,\langle F_3\rangle$ are distinct hyperplanes of $\langle C\rangle$ with the common hyperplane $\langle F\rangle$. Choose $x_k \in F_k \setminus F$ $(k = 1,2,3)$. Since $C \cap \langle F_k\rangle = F_k$ and $C \setminus F_k$ is convex, we have

$$[x_2,x_3] \cap \langle F_1\rangle = \emptyset, \quad [x_1,x_3] \cap \langle F_2\rangle = \emptyset, \quad [x_1,x_2] \cap \langle F_3\rangle = \emptyset.$$

But this contradicts Proposition III.13. □

2 POLYTOPES

We define a *polytope* to be the convex hull of a finite set. Although this definition makes sense in any convex geometry, or even in any aligned space, the theory of polytopes is richer in a linear geometry. We show first that, for polytopes, the notion of 'face' is especially significant.

PROPOSITION 8 *Let P be a polytope and S the (finite) set of extreme points of P. Then the faces of P are the sets $[T]$, where $T \subseteq S$ and $[S \setminus T] \cap \langle T \rangle = \emptyset$.*

Proof By Proposition I.11, $P = [S]$. We show first that if F is a face of P, then $F = [T]$ for some $T \subseteq S$ such that $[S \setminus T] \cap \langle T \rangle = \emptyset$. Evidently we may assume that $F \neq \emptyset, P$. Let T denote the set of all extreme points of P which are contained in F. Then $[S \setminus T] \subseteq P \setminus F$, since $P \setminus F$ is convex, and thus T is a nonempty proper subset of S. If $x \in F$, then $x \in [y,z]$ for some $y \in [S \setminus T]$ and $z \in [T]$. Thus $y \in P \setminus F$ and $z \in F$. Since $x \neq z$ would imply $y \in \langle F \rangle$, which contradicts $P \cap \langle F \rangle = F$, we must have $x = z \in [T]$. Hence $F = [T]$ and $[S \setminus T] \cap \langle T \rangle = \emptyset$.

We show next that $[T]$ is a face of P if T is a subset of S such that $[S \setminus T] \cap \langle T \rangle = \emptyset$. Let $x \in P \cap \langle T \rangle$. Then $x \in [y,z]$, where $y \in [S \setminus T]$ and $z \in [T]$. In fact $x = z$, since $x \neq z$ would imply $y \in \langle T \rangle$. Thus $P \cap \langle T \rangle = [T]$. It remains to show that $P \setminus [T]$ is convex.

Suppose $z \in (x,y)$, where $x \in [S]$, $y \in [S \setminus T]$ and $z \in [T]$. Then $x \in [y',z']$, where $y' \in [S \setminus T]$ and $z' \in [T]$. Since $[S \setminus T] \cap \langle T \rangle = \emptyset$, we must actually have $x \in (y',z')$ and x,y,z' are not collinear. Hence there exists a point $z'' \in (y,y')$ such that $z \in (z',z'')$. Since $z'' \in [S \setminus T] \cap \langle T \rangle$, this is a contradiction.

If $P \setminus [T]$ is not convex then, for some points $x',x'' \in P \setminus [T]$, there exists a point $z \in (x',x'') \cap [T]$. Then $x' \in [y',z']$, where $y' \in [S \setminus T]$ and $z' \in [T]$. Moreover, by what we have already proved, $x' \in (y',z')$ and x'',y',z' are not collinear. Hence there exists a point $z'' \in (x'',y')$ such that $z \in (z',z'')$. Since $z'' \in P \cap \langle T \rangle = [T]$ this yields a contradiction, as we have already seen. □

COROLLARY 9 *If P is a polytope and F a face of P, then F is a polytope and* $E(F) = E(P) \cap F$. □

COROLLARY 10 *A polytope has only finitely many faces. More precisely, a polytope with n extreme points has at most 2^n faces.* □

The bound 2^n in Corollary 10 is actually attained in the case of a *simplex*:

PROPOSITION 11 *Let $P = [S]$, where S is a finite affine independent set. Then a set F is a face of P if and only if $F = [T]$ for some $T \subseteq S$.*

Proof To show that every set $[T]$, where $T \subseteq S$, is a face of P it is sufficient to show that $[S \setminus s]$ is a face of P, for every $s \in S$. But this follows at once from Proposition 8. □

We establish next some further properties, showing that the polytopes considered here behave in the manner to which we are accustomed.

PROPOSITION 12 *If P is a polytope and ℓ a line such that $P \cap \ell \neq \emptyset$, then $P \cap \ell$ is a segment.*

Proof Suppose $P = [S]$, where $S = \{s_1,...,s_n\}$. Since the result is obvious if $n \leq 2$, we assume that $n > 2$ and the result holds for all smaller values of n. Obviously we may assume also that ℓ contains two distinct points x,y of P. The points of the line ℓ may be totally ordered, as in Proposition III.4. By the induction hypothesis, $[S \setminus s_i] \cap \ell$ either is empty or has the form $[a_i,b_i]$, with $a_i \leq b_i$, for each $i \in \{1,...,n\}$. If we put $a = \min_i a_i$ and $b = \max_i b_i$, then $[a,b] \subseteq P \cap \ell$. Moreover, by Proposition II.16, P cannot contain a point of $\ell \setminus [a,b]$. □

It may be noted that the axiom (P) is not required in the proof of Proposition 12.

PROPOSITION 13 *If P is a polytope and L an affine set, then $P \cap L$ is a polytope.*

Proof Suppose $P = [S]$, where $S = \{x_1,...,x_n\}$. Since the result is obvious if $n = 1$, we assume that $n > 1$ and the result holds for all smaller values of n. Let $p \in P \cap L$. By Proposition II.14, we have $P = \bigcup_{i=1}^n P_i$, where $P_i = [p \cup (S \setminus x_i)]$. It is enough to show that the sets $P_i \cap L$ $(i = 1,...,n)$ are all polytopes. For if $P_i \cap L$ is the convex hull of a finite set S_i $(i = 1,...,n)$, then

$$P \cap L = \bigcup_{i=1}^n P_i \cap L \subseteq [S_1 \cup ... \cup S_n].$$

Since $S_1 \cup ... \cup S_n \subseteq P \cap L$ and $P \cap L$ is convex, we must in fact have $[S_1 \cup ... \cup S_n] = P \cap L$.

Without loss of generality, we show only that the set $P_1 \cap L$ is a polytope. Put $P_1' = [x_2,...,x_n]$ and $P' = P_1' \cap L$. Since P' is the convex hull of a finite set, by the induction hypothesis, we will complete the proof by showing that

$$P_1 \cap L = [p \cup P'].$$

The right side is certainly contained in the left, since $p \cup P' = (p \cup P_1') \cap L$ and $P_1 = [p \cup P_1']$. On the other hand, the left side is contained in the right. For if $x \in P_1 \setminus p$ then $x \in (p,y]$ for some $y \in P_1'$, and if $x \in (P_1 \cap L) \setminus p$ then $y \in P_1' \cap L$. □

PROPOSITION 14 *The intersection of a polytope P with a closed half-space of X is again a polytope.*

Proof Let H_+ be an open half-space of X associated with a hyperplane H and $H_+ \cup H$ the corresponding closed half-space. To show that $P \cap (H_+ \cup H)$ is a polytope we may obviously assume that P is not contained in $H_+ \cup H$ and, by Proposition 13, we may assume that $P \cap (H_+ \cup H)$ is not contained in H. Then $X \setminus H$ is not convex and P contains points in both open half-spaces H_+, H_- of X associated with the hyperplane H. Hence the sets S_+, S_- of extreme points of P in H_+, H_- are both nonempty. If S' is the set of extreme points of the polytope $P' = P \cap H$, then the intersections of P with the closed half-spaces $H_+ \cup H$, $H_- \cup H$ are the polytopes $P_+ = [S' \cup S_+]$, $P_- = [S' \cup S_-]$, since $P_+ \cup P_- = P$, $P_+ \cap P_- = P'$. □

PROPOSITION 15 *Any proper face F of a polytope P is contained in a facet of P. Moreover, if* dim F = dim $P - 2$, *then F is contained in exactly two facets of P and is their intersection.*

Proof We begin by proving the following assertion:

(#) Let S be the set of extreme points of P, let S_1 be a subset of S such that $L_1 := \langle S_1 \rangle \subset \langle S \rangle$ and let F_1 be a facet of the polytope $P_1 = P \cap L_1$. If $L_2 := \langle b, S_1 \rangle$, where $b \in S \setminus L_1$, then the polytope $P_2 = P \cap L_2$ has a facet F_2 such that $F_1 \subset F_2$.

By hypothesis $H_1 = \langle F_1 \rangle$ is a hyperplane of L_1 and, if $a \in S_1 \setminus H_1$, P_1 is contained in the closed half-space of L_1 associated with H_1 which contains a. Let T be the set of extreme points of P_2 and let M be the open half-space of L_2 associated with the hyperplane L_1 which contains b. We can choose $b_1,\dots,b_m \in T \cap M$ so that every element of $T \cap M$ belongs to $\langle b_i,H_1\rangle$ for some i and $b_j \notin \langle b_i,H_1\rangle$ if $j \neq i$. We will show that the set $\{b_1,\dots,b_m\}$ is partially ordered by writing $b_i \leq b_j$ if $[a,b_j] \cap \langle b_i,H_1\rangle \neq \emptyset$.

It is evident that $b_i \leq b_i$ for every i. Suppose $b_i \leq b_j$ and $b_j \leq b_i$. Then there exist points $c_i \in [a,b_j] \cap \langle b_i,H_1\rangle$ and $c_j \in [a,b_i] \cap \langle b_j,H_1\rangle$. Hence there exist a point $x \in [b_i,c_i] \cap [b_j,c_j]$ and a point $y \in [b_i,b_j]$ such that $x \in [a,y]$. Then $x \in \langle b_i,H_1\rangle \cap \langle b_j,H_1\rangle$. Moreover $x \notin H_1$, since $y \notin L_1$. Hence $\langle b_i,H_1\rangle = \langle x,H_1\rangle = \langle b_j,H_1\rangle$, which implies $b_i = b_j$.

Suppose next that $b_i \leq b_j$ and $b_j \leq b_k$. We wish to show that $b_i \leq b_k$. There exist points $c_i \in [a,b_j] \cap \langle b_i,H_1\rangle$ and $c_j \in [a,b_k] \cap \langle b_j,H_1\rangle$. If there exists a point $x \in [b_j,c_j] \cap \langle b_i,H_1\rangle$, then $x \in [a,y]$ for some $y \in [b_j,b_k]$. Moreover $x \notin H_1$, since $y \notin L_1$, and hence, as before, $b_i = b_j \leq b_k$. Thus we may now assume $[b_j,c_j] \cap \langle b_i,H_1\rangle = \emptyset$. Then b_j and c_j lie in the same open half-space of L_2 associated with the hyperplane $\langle b_i,H_1\rangle$. Since a and b_j lie in different open half-spaces, it follows that a and c_j lie in different open half-spaces, and hence so also do a and b_k. Thus $[a,b_k] \cap \langle b_i,H_1\rangle \neq \emptyset$ and $b_i \leq b_k$.

This completes the proof that the set $\{b_1,\dots,b_m\}$ is partially ordered. We now choose the notation so that b_m is a maximal element of this partially ordered set. Thus, putting $H_2 = \langle b_m,H_1\rangle$, we have $[a,b_i] \cap H_2 = \emptyset$ for all $i < m$. We are going to show that $F_2 = P \cap H_2$ is a facet of the polytope $P_2 = P \cap L_2$. Since $\langle F_2\rangle = H_2$ is a hyperplane of L_2, we need only show that P_2 is contained in the closed half-space of L_2 associated with the hyperplane H_2 which contains a.

Assume on the contrary that, for some $a' \in T$, there exists a point $x \in (a,a') \cap H_2$. Then, by construction, $a' \notin M$. If $a' \in L_1$ then $x \in H_1$, since $b_m \notin L_1$. But then a and a' lie in different open half-spaces of L_1 associated with the hyperplane H_1, which is a contradiction because F_1 is a facet of P_1. Thus we now suppose $a' \notin L_1$. Then $a' \in N$, where N is an open half-space of L_2 associated with the hyperplane L_1 and $N \neq M$. Thus there exists a point $y \in (a',b_m) \cap L_1$. Hence a,a',b_m are not collinear and there exists a point $z \in (a,y) \cap (b_m,x)$. Since $z \in L_1 \cap H_2$, we must actually have $z \in H_1$. Since $y \in P_1$, this again contradicts the fact that F_1 is a facet of P_1. This completes the proof that F_2 is a facet of P_2, and it is obvious that $F_1 \subset F_2$. Thus we have now proved (#).

If $F_2' = P \cap L_1$ is a facet of P_2, then $F_2 \cap F_2' = F_1$ since $H_2 \cap L_1 = H_1$. If $P \cap L_1$ is not a facet of P_2, then $T \cap N \neq \emptyset$ and we can in the same way choose points $d_1,\ldots,d_n \in T \cap N$ and show that $F_2' = P \cap H_2'$, where $H_2' = \langle d_n, H_1 \rangle$, is also a facet of P_2 containing F_1. Since the segment (b_m,d_n) contains a point $x \in L_1$, and $H_2' = H_2$ implies $x \in H_1$, the facets F_2 and F_2' will be distinct if $(b_m,d_n) \cap H_1 = \emptyset$. Thus they are certainly distinct if F_1 is a face of P_2. Again, $F_2 \cap F_2' = F_1$ since $H_2 \cap H_2' = H_1$. If we take $F_1 = F$ to be a face of P such that $\dim F = \dim P - 2$, then $P_2 = P$ and F is the intersection of the facets F_2, F_2' of P. Since F is contained in at most two facets of P, by Proposition 7, this completes the proof of the second statement of the proposition.

To prove the first statement of the proposition, take $F_1 = F$ to be a face of P such that $\dim F \leq \dim P - 2$ and put $H_1 = \langle F_1 \rangle$. Then $H_1 = \langle R \rangle$, where R is the set of extreme points of P which are contained in F. If $L_1 = \langle a,R \rangle$, where $a \in S \setminus H_1$, then F_1 is a facet of $P_1 = P \cap L_1$. By repeatedly applying (#), we see that F is contained in a facet of P. □

From Proposition 15 we can deduce many other properties of polytopes:

COROLLARY 16 *If F and G are faces of a polytope P such that $F \subset G$, then there exists a finite sequence $F_0,F_1,\ldots,F_r$ of faces of P, with $F_0 = F$ and $F_r = G$, such that F_{i-1} is a facet of F_i $(i = 1,\ldots,r)$.* □

PROPOSITION 17 *Any proper face F of a polytope P is the intersection of all facets of P which contain it.*

Proof Put $d = \dim P - \dim F$. If $d = 1$ the result is obvious and if $d = 2$ it holds by Proposition 15. We assume that $d > 2$ and the result holds for all smaller values of d. There exist faces G,G' of P such that $G \cap G' = F$ and F is a facet of both G and G'. Since, by the induction hypothesis, G and G' are the intersections of all facets of P which contain them, so also is F. □

PROPOSITION 18 *If F,G,H are faces of a polytope P such that $F \subseteq G \subseteq H$, then there exists a face G' of P such that $F \subseteq G' \subseteq H$, $F = G \cap G'$, and every face of P which contains both G and G' also contains H.*

Proof We may suppose $F \subset G \subset H$, since if $G = F$ we can take $G' = H$ and if $G = H$ we can take $G' = F$. Put $n = \dim H - \dim F$. If $n = 2$ the result holds by Proposition 15. We assume that $n > 2$ and the result holds for all smaller values of

n. If $m = \dim G - \dim F$ then, since $n > 2$, either $m > 1$ or $m < n - 1$ (or both). Without loss of generality suppose $m > 1$ and let F_1 be a face of P such that $F \subset F_1 \subset G$ and F is a facet of F_1. Then by the induction hypothesis there is a face G'' of P such that $F_1 \subseteq G'' \subseteq H$, $F_1 = G \cap G''$ and every face which contains both G and G'' also contains H. Since $G'' \neq H$, there also exists a face G' of P such that $F \subseteq G' \subseteq G''$, $F = F_1 \cap G'$ and every face which contains both F_1 and G' also contains G''. Since $G' \subseteq G''$,

$$G \cap G' = (G \cap G'') \cap G' = F_1 \cap G' = F$$

and, since $F_1 \subseteq G$, any face which contains both G and G' also contains G'' and hence also contains H. □

The next result shows, in particular, that polytopes are *polyhedra*:

PROPOSITION 19 *Let P be a polytope and $F_1,\ldots,F_m$ its facets. If Q_i is the closed half-space of $\langle P\rangle$ associated with the hyperplane $H_i = \langle F_i\rangle$ which contains P ($i = 1,\ldots,m$), then*

$$P = \bigcap_{i=1}^{m} Q_i .$$

Proof Put $Q = \bigcap_{i=1}^{m} Q_i$. Then obviously $P \subseteq Q$. We will assume that there exists a point $a \in Q \setminus P$ and derive a contradiction. Since $a \in \langle P\rangle$, we have $a \in \langle x,y\rangle$ for some distinct points $x,y \in P$, by Proposition III.7. Moreover, by Proposition 12, we may choose x,y so that $P \cap \langle x,y\rangle = [x,y]$ and we may choose the notation so that $x \in (a,y)$. Then x and y belong to proper faces of P, by Corollary 6. Hence, by Proposition 15, x and y belong to facets of P. Thus $x \in H_k$ for some k. Since a and y cannot lie in different open half-spaces of $\langle P\rangle$ associated with the hyperplane H_k, it follows that $a,y \in H_k$ and hence y belongs to the same facets of P as x. Hence, by Proposition 17, y belongs to every face of P which contains x. In particular, by Corollary 6 again, $x \in (y,y')$ for some $y' \in P$. But this is a contradiction. □

From Propositions 14 and 19 we immediately obtain

PROPOSITION 20 *The intersection of two polytopes is again a polytope.* □

We can now show also that two-dimensional polytopes are *polygons*:

PROPOSITION 21 *If P is a two-dimensional polytope, then the $n \geq 3$ extreme points of P can be numbered $e_1,\ldots,e_n$ so that the facets of P are precisely the segments $[e_i,e_{i+1}]$ $(i = 1,\ldots,n-1)$ and $[e_n,e_1]$.*

Proof Since the result is obvious if $n = 3$, we assume that $n > 3$ and the result holds for two-dimensional polytopes with fewer than n extreme points. Let S denote the set of extreme points of P. If $e \in S$ then, by Proposition 15, there exist distinct $e',e'' \in S$ such that $[e,e']$ and $[e,e'']$ are facets of P. Suppose $f \in S$ and $f \neq e,e',e''$. Then $\langle e,e'\rangle \cap [e'',f] = \emptyset = \langle e,e''\rangle \cap [e',f]$. Hence, by Lemma III.9, $\langle e,f\rangle \cap [e',e''] \neq \emptyset$. Since all points involved are extreme points of P, we must actually have $(e,f) \cap (e',e'') \neq \emptyset$. It follows that $[e',e'']$ is a facet of the polytope $[S \setminus e]$. The result now follows from the induction hypothesis. □

With any polytope P there is associated a *graph*, whose vertices are the extreme points of P and whose edges are the edges of P. It will now be shown that the graph of a d-dimensional polytope is *d-connected*, i.e. it is connected and remains so whenever less than d vertices are removed. This was first proved, for polytopes in Euclidean space, by Balinski (1961).

LEMMA 22 *Let P be a polytope, G a facet of P and F a face of G. Then there exists an extreme point e' of P such that $e' \notin \langle G\rangle$ and $P \cap \langle e',F\rangle$ is a face of P.*

Proof Put $d = \dim P$. Since the result is trivial when $d = 0$ or 1, we assume that $d \geq 2$ and the result holds for all smaller values of d.

We may suppose $F \subset G$, since if $F = G$ we can take e' to be any extreme point of P which is not in G. Let F' be a facet of G which contains F, and let G' be a facet of P such that $F' = G \cap G'$. By the induction hypothesis there exists an extreme point e' of G' such that $e' \notin \langle F'\rangle$ and $G' \cap \langle e',F\rangle$ is a face of G'. Then e' is an extreme point of P, $e' \notin \langle G\rangle$ and $P \cap \langle e',F\rangle = G' \cap \langle e',F\rangle$ is a face of P. □

PROPOSITION 23 *Let P be a d-dimensional polytope, S the set of extreme points of P and E a subset of S such that $0 \leq |S \setminus E| < d$. Then any two distinct points $e,e' \in E$ can be connected in E by edges of P, i.e. there exists a finite sequence $e_0,e_1,\ldots,e_m$ of elements of E, with $e_0 = e$ and $e_m = e'$, such that $[e_{i-1},e_i]$ is an edge of P for $i = 1,\ldots,m$.*

Proof Fix $e \in E$. It follows from Lemma 22, with $F = \{e\}$, that there exists an affine independent set V of $d+1$ extreme points of P, including e, such that $[e,f]$ is an edge of P for every $f \in V \setminus e$. Hence we may assume that $e' \notin V$ and that the result holds for all d-dimensional polytopes with fewer extreme points than P. Since $\langle V \rangle = \langle P \rangle$, it follows that $e' \in \langle S \setminus e' \rangle$.

By applying Proposition 19 to the polytope $Q = [S \setminus e']$, we see that there exists a facet F of Q such that e' does not lie in the closed half-space of $\langle P \rangle = \langle Q \rangle$ associated with the hyperplane $\langle F \rangle$ which contains Q. We have $F = [T]$, where $T \subset S \setminus e'$. If $f \in T$ then $[e',f]$ is an edge of the polytope $[e' \cup T]$, by Proposition 8, and hence also of P, by Proposition 4. Thus we may assume that $e \in U$, where $U = S \setminus (e' \cup T)$. By the induction hypothesis any two distinct extreme points of Q which are in E can be connected in E by edges of Q. Hence, by Proposition 4, e can be connected in E to some extreme point $f \in T \cap E$ by edges of P. Consequently e can also be connected to e'. □

3 FACTOR GEOMETRIES

In this section we show how new linear geometries may be created from a given one. This property, and the theory of polytopes constructed in the previous section, are persuasive evidence that the concept of linear geometry has intrinsic significance.

Let X be an arbitrary linear geometry, let H_+ be an open half-space associated with a hyperplane H of X, and let L be any nonempty affine subset of H. For any $a \in H_+$, we denote by ρ_a the closed half-space of $\langle a \cup L \rangle$ associated with the hyperplane L which contains a. In symbols, $\rho_a = \langle a \cup L \rangle \cap (H_+ \cup H)$. Consequently $\rho_a \setminus L = \langle a \cup L \rangle \cap H_+$ and $\langle \rho_a \rangle = \langle a \cup L \rangle$. Evidently $\rho_{a'} = \rho_a$ for any $a' \in \rho_a \setminus L$.

The collection $H_+ : L$ of all such closed half-spaces ρ_a with $a \in H_+$ can be given the structure of a linear geometry, the *factor geometry* of H_+ with respect to L, in the following way. For any $\rho_1, \rho_2 \in H_+ : L$, we define the *sector* $[\rho_1, \rho_2]$ to be the set of all $\rho \in H_+ : L$ such that $a \in [a_1, a_2]$ for some $a \in \rho \setminus L$, $a_1 \in \rho_1 \setminus L$ and $a_2 \in \rho_2 \setminus L$. As we now show, we may actually fix a_1 and a_2.

LEMMA 24 *Suppose* $\rho \in [\rho_1,\rho_2]$, *where* $\rho_1,\rho_2 \in H_+ : L$. *If* $a_1 \in \rho_1 \setminus L$ *and* $a_2 \in \rho_2 \setminus L$, *then there exists* $a \in \rho \setminus L$ *such that* $a \in [a_1,a_2]$.

Proof Evidently we may assume $\rho_1 \neq \rho_2$ and $\rho \neq \rho_1,\rho_2$. Then $<\rho_1> \neq <\rho_2>$. By hypothesis there exist points $a' \in \rho \setminus L$, $a_1' \in \rho_1 \setminus L$ and $a_2' \in \rho_2 \setminus L$ such that $a' \in [a_1',a_2']$. In fact $a' \in (a_1',a_2')$, since $\rho \neq \rho_1,\rho_2$. If we put $L_i = <\rho_i> = <a_i' \cup L>$ $(i = 1,2)$, then L_1 and L_2 are hyperplanes of $L^* = <a_1',a_2',L>$. Moreover $L_1 \cap L_2 = L$, by the exchange property. Since $a' \in (a_1',a_2')$, it follows that $L' = <\rho> = <a' \cup L>$ is also a hyperplane of L^* and $L' \neq L_1,L_2$. Moreover a_1' and a_2' lie in different open half-spaces of L^* associated with the hyperplane L'. From $L_1 \cap L' = L$ and $[a_1,a_1'] \cap L = \emptyset$ we obtain $[a_1,a_1'] \cap L' = \emptyset$. In the same way $[a_2,a_2'] \cap L' = \emptyset$. Hence also a_1 and a_2 lie in different open half-spaces of L^* associated with the hyperplane L'. Consequently there exists a point $a \in (a_1,a_2) \cap L'$. Evidently a and a_2 lie in the same open half-space of L^* associated with the hyperplane L_1. But a and a' also lie in the same open half-space of L^* associated with the hyperplane L_1, since $L_1 \cap L' = L$. Therefore $[a',a_2] \cap L_1 = \emptyset$. Assume $a \notin \rho$. Then there exist a point $x \in (a,a') \cap L$ and a point $y \in (a',a_2)$ such that $x \in (a_1,y)$. Since $y \in [a',a_2] \cap L_1$, this is a contradiction. Thus $a \in \rho$. □

It will now be shown that, with sectors as 'segments', $H_+ : L$ is a linear geometry.

LEMMA 25 *Suppose* $x \in [a,b,c]$, *where* a,b,c *are non-collinear points. If* $x \notin [a,b],[a,c],[b,c]$, *then* $x \in (a,y)$ *for some* $y \in (b,c)$.

Proof By Proposition II.16 there exist points $y,z \in [a,b] \cup [a,c] \cup [b,c]$ such that $a,x \in [y,z]$. Moreover $a \in \{y,z\}$, since a is an extreme point of $[a,b,c]$. Without loss of generality assume $a = z$. Then $y \in (b,c)$, since $x \notin [a,b],[a,c]$, and $x \neq y$, since $x \notin [b,c]$. □

PROPOSITION 26 *If* H_+ *is an open half-space associated with a hyperplane* H *of* X *and* L *any nonempty affine subset of* H, *then* $H_+ : L$ *is a linear geometry.*

Proof We will prove that $H_+ : L$ satisfies each of the axioms **(C)**, **(L1)–(L4)** and **(P)**. In only one case, namely **(L3)**, does the proof present difficulty.

(L1): It is obvious that $[\rho,\rho] = \{\rho\}$ for any $\rho \in H_+ : L$.

(C): Suppose $\rho' \in [\rho,\rho_1]$ and $\rho'' \in [\rho',\rho_2]$. Choose $a \in \rho \setminus L$, $a_1 \in \rho_1 \setminus L$ and $a_2 \in \rho_2 \setminus L$. Then, by Lemma 24, there exist a point $a' \in \rho' \setminus L$ such that $a' \in [a,a_1]$ and a point $a'' \in \rho'' \setminus L$ such that $a'' \in [a',a_2]$. But $a'' \in [a,\tilde{a}]$ for some $\tilde{a} \in [a_1,a_2]$. If $\tilde{a} \notin L$, it follows that $\rho'' \in [\rho,\tilde{\rho}]$ for some $\tilde{\rho} \in [\rho_1,\rho_2]$. On the other hand, if $\tilde{a} \in L$ then $a'' \in \rho \setminus L$ and $\rho'' = \rho$.

This already proves that $H_+ : L$ is a convex geometry.

(P): Suppose $\rho_1' \in [\rho,\rho_1]$ and $\rho_2' \in [\rho,\rho_2]$. Choose $a \in \rho \setminus L$, $a_1 \in \rho_1 \setminus L$ and $a_2 \in \rho_2 \setminus L$. Then there exist a point $a_1' \in \rho_1' \setminus L$ such that $a_1' \in [a,a_1]$ and a point $a_2' \in \rho_2' \setminus L$ such that $a_2' \in [a,a_2]$. Hence there exists a point $a' \in [a_1,a_2'] \cap [a_1',a_2]$. Moreover $a' \in H_+$, since $a_1,a_2' \in H_+$. Consequently $\rho_{a'} \in [\rho_1,\rho_2'] \cap [\rho_1',\rho_2]$.

(L4): Suppose $\rho_3 \in [\rho_1,\rho_2]$ and $\rho \in [\rho_1,\rho_2]$. Choose $a_i \in \rho_i \setminus L$ $(i = 1,2)$. Then there exist $a_3 \in \rho_3 \setminus L$ such that $a_3 \in [a_1,a_2]$ and $a \in \rho \setminus L$ such that $a \in [a_1,a_2]$. Hence $a \in [a_1,a_3] \cup [a_3,a_2]$, and $\rho \in [\rho_1,\rho_3] \cup [\rho_3,\rho_2]$. Thus $[\rho_1,\rho_2] \subseteq [\rho_1,\rho_3] \cup [\rho_3,\rho_2]$.

On the other hand, if $\rho \in [\rho_1,\rho_3]$ there exists $a \in \rho \setminus L$ such that $a \in [a_1,a_3] \subseteq [a_1,a_2]$. Hence $[\rho_1,\rho_3] \subseteq [\rho_1,\rho_2]$, and similarly $[\rho_3,\rho_2] \subseteq [\rho_1,\rho_2]$. Thus $[\rho_1,\rho_2] = [\rho_1,\rho_3] \cup [\rho_3,\rho_2]$.

(L2): Suppose $\rho_2 \in [\rho_1,\rho_3]$, $\rho_3 \in [\rho_2,\rho_4]$ and $\rho_2 \neq \rho_3$. Choose $a_i \in \rho_i \setminus L$ $(i = 1,3,4)$. Then there exist $a_2 \in \rho_2 \setminus L$ such that $a_2 \in [a_1,a_3]$ and $a_3' \in \rho_3 \setminus L$ such that $a_3' \in [a_2,a_4]$. Moreover $a_2 \neq a_3$, since $\rho_2 \neq \rho_3$. If $a_3' = a_3$, it follows that $a_3 \in [a_1,a_4]$ and hence $\rho_3 \in [\rho_1,\rho_4]$. If $a_3' \neq a_3$, then $a_3' \in [a_3,a_3'']$ for some $a_3'' \in [a_1,a_4]$. Since $a_3'' \in \langle\rho_3\rangle \cap H_+$, it follows that $a_3'' \in \rho_3 \setminus L$ and thus again $\rho_3 \in [\rho_1,\rho_4]$.

(L3): Suppose $\rho_3 \notin [\rho_1,\rho_2]$ and $\rho_2 \notin [\rho_1,\rho_3]$. We wish to show that $[\rho_1,\rho_2] \cap [\rho_1,\rho_3] = \{\rho_1\}$. We may suppose $\rho_1 \notin [\rho_2,\rho_3]$, since otherwise the conclusion follows from Proposition II.18 (iii) (whose hypotheses we have already shown to be satisfied). Choose $a_i \in \rho_i \setminus L$ $(i = 1,2,3)$. Then a_1,a_2,a_3 are not collinear and $[a_1,a_2] \cap \langle a_3,L\rangle = \emptyset$, $[a_1,a_3] \cap \langle a_2,L\rangle = \emptyset$, $[a_2,a_3] \cap \langle a_1,L\rangle = \emptyset$. Hence, by Lemma III.9, $\langle a_1,a_2,a_3\rangle \cap L = \emptyset$.

Assume, on the contrary, that there exists $\rho \in [\rho_1,\rho_2] \cap [\rho_1,\rho_3]$ with $\rho \neq \rho_1$. Then there exist $a',a'' \in \rho \setminus L$ such that $a' \in (a_1,a_2)$, $a'' \in (a_1,a_3)$. Moreover

$a' \neq a''$, since a_1,a_2,a_3 are not collinear. Since $a' \in <a'',L>$, it follows from Proposition III.7 that $a' \in <a'',l',l''>$ for some $l',l'' \in L$. Moreover $<a',a''> \cap L = \emptyset$, since $a',a'' \in <a_1,a_2,a_3>$, and hence $l' \neq l''$.

Put $Y = <l',l'',a_1,a_3>$. Since $a'' \in Y$, we must have $a' \in Y$ and hence also $a_2 \in Y$. Since Y contains the affine independent points a_1,a_2,a_3,l', it follows that $\dim Y = 3$. But, since $H_k = <l',l'',a_k>$ $(k = 1,2,3)$ are distinct planes in Y with the common line $Z = <l',l''>$, this contradicts Proposition III.13. □

4 NOTES

The faces of arbitrary convex sets do not share many of the properties possessed by the faces of polytopes (or of polyhedra, which will be considered in Chapter V). For example in $\mathbb{R}^3$, with the usual definition of convexity, let $C = [p,q,S]$ be the convex hull of the set consisting of the circle $S = \{x = (\xi_1,\xi_2,\xi_3): (\xi_1 - 1)^2 + \xi_2^2 = 1, \xi_3 = 0\}$ and the two points $p = (0,0,1)$, $q = (0,0,-1)$. The extreme points of C are p,q and all points of S except the origin. The edges of C are the segments whose endpoints are distinct extreme points, not both of which are in S. However, C has no facets. (It may be noted also that, although C is compact, the set of all extreme points is not closed.)

Polytopes in Euclidean space are extensively treated in the books of Brøndsted (1983), Grünbaum (1967) and Ziegler (1995). The present treatment follows Coppel (1995), but the key Proposition 15 has a stronger formulation here. It is natural to ask if the Euler–Poincaré relation, which connects the number of faces of different dimensions of a polytope in $\mathbb{R}^d$, can be generalized to any linear geometry.

Factor geometries are considered by Prenowitz (1961), Rubinstein (1964) and Prenowitz and Jantosciak (1979), but they assume the axioms **(D)** and **(U)** which will be considered in the next chapter.

V
Density and Unendingness

In this chapter two new axioms are introduced. A linear geometry is *dense* if no segment contains exactly two points. The notions of *intrinsic interior* and *convex closure* of a convex set are defined, and it is shown that in a dense linear geometry they have many basic properties. In particular, any finite-dimensional convex set has a nonempty intrinsic interior. A linear geometry is *unending* if any segment is contained in another segment with different endpoints. An unending linear geometry of dimension greater than one is necessarily also dense. In an unending linear geometry the *affine hull* of the union of two *affine sets* A,B is the union of the affine hulls of all sets $\{a,b\}$, where $a \in A$ and $b \in B$, provided $A \cap B \neq \emptyset$. Furthermore a dense unending linear geometry may be given the structure of a topological space, and actually of a Hausdorff space.

Following Prenowitz, we define *products* of arbitrary nonempty sets in a dense linear geometry, and *quotients* also in a dense unending linear geometry. Products and quotients possess many simple algebraic properties, and their usefulness in elementary geometry has perhaps not been fully appreciated. We further show that a substantial part of the usual theory of *cones* and *polyhedra* in $\mathbb{R}^d$ carries over to any dense unending linear geometry.

1 DENSE LINEAR GEOMETRIES

A linear geometry will be said to be *dense* if it has the following property:

(D) *if $a,b \in X$ and $a \neq b$, then $[a,b] \neq \{a,b\}$.*

It is remarkable that this axiom actually implies the axiom **(P)**, when all the other axioms of a linear geometry are satisfied:

PROPOSITION 1 *A convex geometry is a linear geometry if it satisfies the axioms* **(L2)–(L4)** *and* **(D)**.

Proof It is sufficient to show that $[y_1,z_2] \cap [y_2,z_1] \neq \emptyset$ if $z_1 \in (x,y_1)$, $z_2 \in (x,y_2)$ and $z_1 \neq z_2$. Moreover, by Proposition III.6, we may assume that x,y_1,y_2 are not collinear.

By **(D)** we can choose $z \in (z_1,z_2)$. Then $z \in (x,u)$ for some $u \in (y_2,z_1)$, and similarly $z \in (x,v)$ for some $v \in (y_1,z_2)$. We may assume $u \neq v$, since otherwise there is nothing more to do. Then either $u \in (z,v)$ or $v \in (z,u)$. Without loss of generality suppose $u \in (z,v)$. Then $u \in (z_1,w)$ for some $w \in (y_1,z_2)$ and $w \in (y_2,z_1')$ for some $z_1' \in (x,y_1)$. Since the line $\langle y_2,u\rangle = \langle y_2,w\rangle$ has at most one point in common with the segment (x,y_1), we must have $z_1' = z_1$. Thus $w \in (y_1,z_2) \cap (y_2,z_1)$. □

By Corollary III.15, a linear geometry is dense if not all points are collinear and every line contains at least three points. Furthermore, a dense linear geometry is irreducible. *Throughout the remainder of this section we will assume that a set X is given on which a dense linear geometry is defined.*

PROPOSITION 2 *The class of Carathéodory sets coincides with the class of nonempty finite affine independent sets, and hence also with the class of nonempty finite Helly sets and the class of nonempty finite Radon sets.*

Proof We need only show that every nonempty finite affine independent set is a Carathéodory set, by Propositions II.15 and III.17. Let $S = \{s_1,\ldots,s_n\}$ be an affine independent set. By **(D)** we can choose $x_1 \in (s_1,s_2)$ and then, inductively, $x_k \in (x_{k-1},s_{k+1})$ $(k = 2,\ldots,n-1)$. Evidently $x_{n-1} \in [S]$ and, for each $j < n$, $x_{j-1} \in [s_1,\ldots,s_j] \subseteq \langle S\setminus s_{j+1}\rangle$. However, $x_j \notin \langle S\setminus s_{j+1}\rangle$, since $x_j \in \langle S\setminus s_{j+1}\rangle$ would imply $s_{j+1} \in \langle x_{j-1},x_j\rangle \subseteq \langle S\setminus s_{j+1}\rangle$, which contradicts the affine independence of S. In particular, $x_{n-1} \notin [S\setminus s_n]$.

Assume, contrary to the proposition, that $[S] = \bigcup_{i=1}^{n}[S\setminus s_i]$. Then $x_{n-1} \in [S\setminus s_m]$ for some $m < n$. But if for some k with $m \leq k < n$ we have $x_k \in \langle S\setminus s_m\rangle$ then, since $s_{k+1} \in S\setminus s_m$, we also have $x_{k-1} \in \langle x_k,s_{k+1}\rangle \subseteq \langle S\setminus s_m\rangle$. It follows that $x_m \in \langle S\setminus s_m\rangle$ and, if $m > 1$, even $x_{m-1} \in \langle S\setminus s_m\rangle$. Since the latter is a contradiction, we conclude that $m = 1$ and $x_1 \in \langle S\setminus s_1\rangle$. But this is also a contradiction, since $s_1 \in \langle x_1,s_2\rangle \subseteq \langle S\setminus s_1\rangle$. □

The axiom (**D**) is essential for the validity of Proposition 2. For if x,y are distinct points of a linear geometry such that $[x,y] = \{x,y\}$, then $\{x,y\}$ is an affine independent set but not a Carathéodory set.

PROPOSITION 3 *If C is a convex set and $A_1,...,A_m$ affine sets such that*

$$C \subseteq A_1 \cup ... \cup A_m,$$

then $C \subseteq A_i$ for some $i \in \{1,...,m\}$.

Proof The result is obviously true if $m = 1$. We assume that $m > 1$ and the result holds for all smaller values of m. If C is not contained in A_i for some i then, by the induction hypothesis, there exist a point $a \in C \setminus (\bigcup_{i=2}^{m} A_i)$ and a point $b \in C \setminus (\bigcup_{i=1}^{m-1} A_i)$. Then $a \in A_1$ and $b \in A_m$. If $x \in (a,b)$, then $x \notin A_1 \cup A_m$. Hence $m > 2$ and $x \in A_2 \cup ... \cup A_{m-1}$. Thus if we fix $c \in (a,b)$, then $[a,c] \subseteq A_1 \cup ... \cup A_{m-1}$. Hence, by the induction hypothesis, $[a,c] \subseteq A_1$. Thus $c \in A_1$, which is a contradiction. □

We define the *intrinsic interior* C^i of a convex set C to be the subset of all points of C which do not belong to a proper face of C. Thus $C^i = C$ if C contains at most one point.

PROPOSITION 4 *Let C be a convex set and C^i its intrinsic interior. If $x \in C^i$ and $y \in C$, then $(x,y) \subseteq C^i$. In particular, C^i is also a convex set.*

Proof Let $z \in (x,y)$. If A is a face of C which contains z, then also $x \in A$, by Proposition II.6. Hence $A = C$. □

The definition of 'intrinsic interior' may also be reformulated in the following way:

PROPOSITION 5 *The intrinsic interior C^i of a convex set C is the set of all points $a \in C$ such that, for every $b \in C \setminus a$, there exists some $c \in C$ for which $a \in (b,c)$.*

Proof This follows at once from Corollary IV.6. □

We now introduce another concept of a topological nature and define the *convex closure* $\overline{C}$ of a convex set C to be C itself if $C^i = \emptyset$, and otherwise to be the set of all $y \in X$ such that $[x,y) \subseteq C^i$ for some $x \in C^i$. In either event $C \subseteq \overline{C}$, by Proposition 4.

In the preceding definition of 'convex closure' we can replace 'some' by 'every'. For suppose $[x',y) \subseteq C^i$ and $x \in C^i$. We wish to show that also $[x,y) \subseteq C^i$. We may assume that x,x',y are not collinear, by Proposition 4. Since $x \in C^i$, there exists $x'' \in C$ such that $x \in (x',x'')$. If $z \in (x,y)$, then $z \in (x'',z')$ for some $z' \in (x',y)$. Since $z' \in C^i$ and $x'' \in C$, it follows that $z \in C^i$.

PROPOSITION 6 *If C is a convex set, then $\overline{C}$ is also a convex set.*

Proof Obviously we may assume that $C^i \neq \emptyset$. Suppose $x,y \in \overline{C}$, where $x \neq y$. We wish to show that if $z \in (x,y)$, then also $z \in \overline{C}$. That is, we wish to show that $c \in C^i$ and $w \in (c,z)$ imply $w \in C^i$. Evidently we may suppose that $z \notin [c,x]$ and $z \notin [c,y]$. Then c,x,y are non-collinear and hence, by Proposition II.2′, $w \in (d,y)$ for some $d \in (c,x)$. Then $d \in C^i$, and hence $w \in C^i$. □

The definitions of 'intrinsic interior' and 'convex closure' make sense, but have little significance, in an arbitrary convex geometry and the preceding properties hold in an arbitrary linear geometry. However, the following results will make essential use of the axiom **(D)**.

PROPOSITION 7 *If C is a convex set, then $\overline{C}^i = C^i$ and $\overline{C} = \overline{\overline{C}}$.*

Proof Obviously we may assume that $\overline{C} \neq C$, and then $C^i \neq \emptyset$. Suppose $x \in \overline{C}^i$. Then for every $y \in \overline{C} \setminus x$ there exists $z \in \overline{C}$ such that $x \in (y,z)$. In particular this holds for every $y \in C^i \setminus x$ and then $(y,z) \subseteq C^i$, by the definition of $\overline{C}$. Thus $x \in C^i$ and $\overline{C}^i \subseteq C^i$.

Suppose, on the other hand, that $x \in C^i$ and $y \in \overline{C} \setminus x$. Then $(x,y) \subseteq C^i$ and, by **(D)**, there exists some $w \in (x,y)$. By the definition of C^i there exists $z \in C$ such that $x \in (w,z)$. Then $x \in (y,z)$, by **(L2)**. Thus $x \in \overline{C}^i$ and $C^i \subseteq \overline{C}^i$.

This proves the first assertion of the proposition. To prove the second we need only show that $\overline{\overline{C}} \subseteq \overline{C}$, since the reverse inclusion is trivial. Suppose $x \in \overline{\overline{C}}$. Then $(x,y] \subseteq \overline{C}^i$ for every $y \in \overline{C}^i$. Since $\overline{C}^i = C^i$, by what we have just proved, this implies $x \in \overline{C}$. □

PROPOSITION 8 *If C is a convex set, then $(C^i)^i = C^i$. Moreover, if $C^i \neq \emptyset$ then $\overline{C^i} = \overline{C}$.*

Proof To prove the first assertion we may assume that C^i contains more than one point. Suppose $x,y \in C^i$, where $x \neq y$. Then there exists $z \in C$ such that $x \in (y,z)$. Moreover, by **(D)**, there exists some $w \in (x,z)$. Then $w \in C^i$ and, by Proposition III.1(i), $x \in [y,w]$. Hence $w \neq y$ and $x \in (y,w)$. Thus $x \in C^{ii}$ and $C^{ii} = C^i$.

Since $C^i \neq \emptyset$, $u \in \overline{C}$ if and only if $(u,v] \subseteq C^i$ for every $v \in C^i$. Since $C^{ii} = C^i$, $u \in \overline{C^i}$ if and only if the same condition is satisfied. □

PROPOSITION 9 *If C is a convex set and $c \in C^i$, then $C \cap (c,x) \neq \emptyset$ for each $x \in \langle C\rangle \setminus c$.*

Proof Evidently we may assume that $x \notin C$. By Proposition III.7 we have $x \in \langle c_1,c_2\rangle$ for some $c_1,c_2 \in C$. Since $x \notin [c_1,c_2]$, we may assume the notation chosen so that $c_1 \in (c_2,x)$. Since $c \in C^i$, there exists a point $c_3 \in C$ such that $c \in (c_2,c_3)$. If c,c_1,c_2 are collinear and $c_1 \notin (c,x)$, then $c \in (c_2,x)$ and $c_3 \in (c,x)$. Thus we now assume that c,c_1,c_2 are not collinear. Then, by **(P)′**, there exists a point $c_4 \in (c_1,c_3) \cap (c,x)$. □

PROPOSITION 10 *Let C and D be convex sets such that $C \subseteq D$. If $C \cap D^i \neq \emptyset$, or if $\langle C\rangle = \langle D\rangle$, then $C^i \subseteq D^i$ and $\overline{C} \subseteq \overline{D}$.*

Proof It is enough to show that $C^i \subseteq D^i$, since the relation $\overline{C} \subseteq \overline{D}$ then follows from the definition of convex closure. Obviously we may assume that $C^i \neq \emptyset$.

Let $x \in C^i$ and suppose first that there exists a point $y \in C \cap D^i$, where $y \neq x$. Since $y \in C$, there exists a point $z \in C$ such that $x \in (y,z)$. Since $z \in D$ and $y \in D^i$, it follows that $x \in D^i$.

Suppose next that $\langle C\rangle = \langle D\rangle$ and let $y \in D \setminus C$. Then there exists a point $x' \in C \cap (x,y)$, by Proposition 9, and there exists a point $y' \in C$ such that $x \in (x',y')$, since $x \in C^i$. Since $y' \in D$ and $x \in (y,y')$, it follows that $x \in D^i$. □

PROPOSITION 11 *For any convex sets C,D with $C^i \neq \emptyset$, $D^i \neq \emptyset$, the following statements are equivalent*:

(i) $C^i = D^i$,
(ii) $\overline{C} = \overline{D}$,
(iii) $C^i \subseteq D \subseteq \overline{C}$.

Proof It follows at once from Propositions 7 and 8 that (i) and (ii) are equivalent. Also it is obvious that $C^i = D^i$ implies $C^i \subseteq D$, and that $\overline{C} = \overline{D}$ implies $D \subseteq \overline{C}$.

Suppose finally that $C^i \subseteq D \subseteq \overline{C}$. Applying Proposition 10 to the sets D and $\overline{C}$ we obtain $D^i \subseteq C^i$, and now applying it to the sets C^i and D we obtain $C^i \subseteq D^i$. □

PROPOSITION 12 *If $C = A \cap B$, where A and B are convex sets such that $A^i \cap B^i \neq \emptyset$, then*

$$C^i = A^i \cap B^i, \quad \overline{C} = \overline{A} \cap \overline{B}.$$

Proof Let $z \in A^i \cap B^i$ and $c \in C \setminus z$. Since $z \in A^i$ and $c \in A$, there exists $a \in A$ such that $z \in (a,c)$. Similarly, since $z \in B^i$ and $c \in B$, there exists $b \in B$ such that $z \in (b,c)$. By **(L3)** either $a \in [b,c]$ or $b \in [a,c]$. Hence either $a \in C$ or $b \in C$. In any event, $z \in C^i$. Thus $A^i \cap B^i \subseteq C^i$.

On the other hand, if $w \in C^i$ and $w \neq z$, then there exists $c' \in C$ such that $w \in (z,c')$. It follows that $w \in A^i \cap B^i$. Thus $C^i \subseteq A^i \cap B^i$.

This proves the first statement of the proposition, and the second statement is an immediate consequence. □

We can now derive a converse to Proposition I.7:

PROPOSITION 13 *Let C_1, C_2 be convex sets and let F be a face of $C_1 \cap C_2$ such that $F^i \neq \emptyset$. Then there exist faces F_1 of C_1 and F_2 of C_2, with $F_1{}^i \neq \emptyset$, $F_2{}^i \neq \emptyset$, such that $F = F_1 \cap F_2$.*

Proof Let $x \in F^i$. Then F is the intersection of all faces of $C_1 \cap C_2$ which contain x, since no face of $C_1 \cap C_2$ containing x is properly contained in F and since the intersection of two faces is again a face. Since faces form an alignment, there is a face F_k of C_k containing x which is contained in every face of C_k containing x $(k = 1,2)$. It follows that $x \in F_1{}^i \cap F_2{}^i$, since any face of F_k is also a face of C_k $(k = 1,2)$. Moreover $F \subseteq F_1 \cap F_2$, since $F_1 \cap F_2$ is a face of $C_1 \cap C_2$ which contains x. But x does not belong to any proper face of $F_1 \cap F_2$, since $x \in (F_1 \cap F_2)^i$ by Proposition 12. Since F is a face of $F_1 \cap F_2$, we must have $F = F_1 \cap F_2$. □

We next derive some properties of polytopes:

PROPOSITION 14 *If P is a polytope, then $\overline{P} = P$.*

Proof We may assume that $P^i \neq \emptyset$. Let $x \in P^i$ and assume that there exists a point $y \in \overline{P} \setminus P$. By Proposition IV.12, there exist $a,b \in P$ such that

$P \cap \langle x,y \rangle = [a,b]$. Then $y \notin [a,b]$, $x \in (a,b)$ and we may choose the notation so that $b \in (x,y)$. Hence $b \in P^i$, which is a contradiction. □

PROPOSITION 15 *If $P = [S]$, where S is a finite affine independent set, then $P^i \neq \emptyset$ and*

$$P^i = [S] \setminus \bigcup_{s \in S} [S \setminus s].$$

Proof The formula for P^i follows from Proposition IV.11 and the definition of intrinsic interior. Since S is a Carathéodory set, by Proposition 2, it now follows that $P^i \neq \emptyset$. □

Proposition 15 has the following important consequence:

PROPOSITION 16 *If D is a nonempty finite-dimensional convex set, then $D^i \neq \emptyset$.*

Proof Let S be a maximal affine independent subset of D and put $C = [S]$. Then $C^i \neq \emptyset$, by Proposition 15. But $\langle C \rangle = \langle D \rangle$ and so, by Proposition 10, $C^i \subseteq D^i$. □

Thus if P is a polytope, then $P^i \neq \emptyset$. However, if X is infinite-dimensional, it necessarily contains a nonempty convex set C with $C^i = \emptyset$. Indeed we can take $C = [S]$, where $S = \{x_1, x_2, \ldots\}$ is any countable affine independent subset of X. For assume $x \in C^i$. Since $C = \bigcup_{n=1}^{\infty} C_n$, where $C_n = [x_1, \ldots, x_n]$, we have $x \in C_m$ for some m. For any $y \in C \setminus x$, there exists $z \in C$ such that $x \in (y,z)$. Since $y,z \in C_n$ for some $n > m$ and since C_m is a face of C_n, by Proposition IV.11, it follows that $y \in C_m$. Hence $C = C_m$, which is a contradiction.

It follows at once from Proposition 4 that $\langle C^i \rangle = \langle C \rangle$ if $C^i \neq \emptyset$, and hence $\langle \overline{C} \rangle = \langle C \rangle$ without exception. From the definitions and from Proposition III.12 we also obtain at once

PROPOSITION 17 *Let H be a hyperplane such that $X \setminus H$ is not convex. If H_+ and H_- are the open half-spaces associated with the hyperplane H, then*

$$H_+^i = H_+, \quad H_-^i = H_-,$$

$$\overline{H_+} = H_+ \cup H, \quad \overline{H_-} = H_- \cup H. \quad □$$

We can now also prove

PROPOSITION 18 *If a and b are distinct points of X, then there exists a hyperplane H such that a and b lie in different open half-spaces associated with H.*

Proof Choose $c \in (a,b)$ and consider the family $\mathcal{F}$ of all affine sets which contain c but not a. If H is the union of a maximal totally ordered subfamily of $\mathcal{F}$, then $H \in \mathcal{F}$. Moreover, if $x \in X \setminus H$ then $a \in \langle H \cup x \rangle$ and hence $x \in \langle H \cup a \rangle$, by Proposition III.8. Thus $\langle H \cup a \rangle = X$ and H is a hyperplane. □

For any $a,b \in X$, put

$$\begin{aligned} ab &= (a,b) \text{ if } a \neq b, \\ &= \{a\} \text{ if } a = b. \end{aligned}$$

According to our definition, ab is not a segment if $a \neq b$, since $a,b \notin ab$. However, we will show that the analogues of the axioms **(C)** and **(P)** are satisfied.

We show first that *if $c \in ab_1$ and $d \in cb_2$, then $d \in ab$ for some $b \in b_1b_2$.* If a,b_1,b_2 are not collinear this follows from Proposition II.2′, and if a,b_1,b_2 are not all distinct it follows readily from **(D)**. Suppose then that a,b_1,b_2 are distinct but collinear. If $a \in (b_1,b_2)$ then either $d = a$ and we can take $b = a$, or $d \in (a,c)$ and we can take any $b \in (c,d)$, or $d \in (a,b_2)$ and we can take any $b \in (d,b_2)$. If $b_2 \in (a,b_1)$ we can take any $b \in (b_1,b_2) \cap (c,b_1)$. If $b_1 \in (a,b_2)$ we can take any $b \in (b_1,b_2) \cap (d,b_2)$.

We show next that *if $c_1 \in ab_1$ and $c_2 \in ab_2$, then $b_1c_2 \cap b_2c_1 \neq \emptyset$.* If a,b_1,b_2 are not collinear this follows from **(P)′**, and if a,b_1,b_2 are not all distinct it follows readily from **(D)**. Suppose then that a,b_1,b_2 are distinct but collinear. If $a \in (b_1,b_2)$, then $a \in (b_1,c_2) \cap (b_2,c_1)$. Suppose $b_2 \in (a,b_1)$. If $c_1 \in (b_1,c_2)$ then $(b_2,c_1) \subseteq (b_1,c_2)$, and if $c_2 \in [c_1,b_1)$ then $(b_2,c_2) \subseteq (b_1,c_2) \cap (b_2,c_1)$. The same argument applies if $b_1 \in (a,b_2)$.

We will refer to these analogues of the axioms **(C)** and **(P)** as **(C)″** and **(P)″**. It is easily verified that the analogues of Propositions II.2, II.17 and II.19 also remain valid, i.e.

(i) *if $c_1 \in ab_1$, $c_2 \in ab_2$ and $c \in c_1c_2$, then $c \in ab$ for some $b \in b_1b_2$;*

(ii) *if $c_1 \in ab_1$, $c_2 \in ab_2$ and $b \in b_1b_2$, then there exists a point $c \in ab \cap c_1c_2$;*

(iii) *if $x \in aa' \cap bb'$ and $y \in ab$, then $x \in yy'$ for some $y' \in a'b'$.*

When required, these analogues will be referred to as Propositions II.2″, II.17″ and II.19″.

If A and B are nonempty subsets of X, we define their *product* AB by

$$AB = \bigcup_{a \in A, b \in B} ab.$$

In other words, AB is the union of $A \cap B$ with the set of all points $c \in X$ such that $c \in (a,b)$ for some distinct points $a \in A$ and $b \in B$. The reason for the terminology and notation is that products possess many of the properties of products of positive numbers, as we now show.

First of all $AB \neq \emptyset$, on account of (D), and it is obvious that, if $A \subseteq C$ and $B \subseteq D$, then $AB \subseteq CD$. Evidently also $AB = BA$. Furthermore the associative law holds: for any nonempty sets A,B,C,

$$(AB)C = A(BC).$$

Indeed if $x \in (AB)C$ then, for some $a \in A$, $b \in B$, $c \in C$, we have $x \in cy$ and $y \in ab$. Hence, by (C)″, $x \in az$ for some $z \in bc$. Thus $(AB)C \subseteq A(BC)$ and, for the same reason,

$$A(BC) = (CB)A \subseteq C(BA) = (AB)C.$$

For any nonempty set $A \subseteq X$ we have $A \subseteq AA$. Moreover it follows at once from the definitions that A is convex if and only if $AA = A$.

The proof of the next proposition illustrates how the algebraic properties of products can replace involved geometric arguments.

PROPOSITION 19 *If A and B are nonempty convex sets, then their product AB is also a nonempty convex set.*

Proof $(AB)(AB) = (AA)(BB) = AB$. □

The following proposition essentially just summarises some earlier results:

PROPOSITION 20 *Let C be a nonempty convex set. Then*

(i) *$x \in C^i$ if and only if $x \in cC$ for every $c \in C$,*
(ii) *if $x \in C^i$, then $xC = C^i$,*
(iii) *if $C^i \neq \emptyset$, then $C^i = C^i c$ for every $c \in C$.*

Proof The statement (i) is just a reformulation of Proposition 5, and (ii) follows at once from (i) and Proposition 4.

If $C^i \neq \varnothing$, then (ii) implies that $C^i c \subseteq C^i$ for any $c \in C$. On the other hand, if $x \in C^i$ and $c \in C$, then $x \in cc'$ for some $c' \in C$. If we choose $x' \in c'x$, then $x' \in C^i$ and $x \in cx'$. Hence $C^i \subseteq C^i c$. This proves (iii). □

In the statements of (i)–(iii) we may replace C by $\overline{C}$ (and keep C^i), by Proposition 7.

PROPOSITION 21 *Let A and B be nonempty convex sets. If $x \in AB$, $y \in [A \cup B]$ and $y \neq x$, then $(x,y) \subseteq AB$.*

Proof Let $z \in (x,y)$. We have $y \in [a'',b'']$ for some $a'' \in A$ and $b'' \in B$. If $y \in AB$ then also $z \in AB$, since AB is convex. Hence we may assume without loss of generality that $y \in A$ and $y \notin B$. Since $x \in ab$ for some $a \in A$ and $b \in B$, it follows that $z \in xy \subseteq yab$ and hence $z \in a'b$ for some $a' \in A$. □

PROPOSITION 22 *If A and B are nonempty convex sets, then*

$$[A \cup B]^i = (AB)^i.$$

Proof We may assume that $[A \cup B]$ is not a singleton, and then also AB is not a singleton by Proposition 21.

We show first that $[A \cup B]^i \subseteq (AB)^i$. If $x \in [A \cup B]^i$, $y \in AB$ and $y \neq x$, then $x \in (y,z)$ for some $z \in [A \cup B]$. If $z' \in (x,z)$, then $z' \in (y,z)$ and hence $z' \in AB$ by Proposition 21. Since $x \in (y,z')$, it follows that $x \in AB$ and actually $x \in (AB)^i$.

We show next that $(AB)^i \subseteq [A \cup B]^i$. Suppose $x \in (AB)^i$, $y \in [A \cup B]$ and $y \neq x$. If $y' \in (x,y)$, then $y' \in AB$ by Proposition 21. Hence $x \in (y',z)$ for some $z \in AB$. Since $AB \subseteq [A \cup B]$ and $x \in (y,z)$, it follows that $x \in [A \cup B]^i$. □

PROPOSITION 23 *If A and B are convex sets such that $A^i \neq \varnothing$ and $B^i \neq \varnothing$, then*

$$[A \cup B]^i = (AB)^i = A^iB^i \neq \varnothing.$$

Proof By Proposition 22 we need only show that $(AB)^i = A^iB^i$. We divide the proof into two parts.

We show first that $(AB)^i \subseteq A^iB^i$. Let $x \in (AB)^i$ and let $y \in A^iB^i$ with $y \neq x$. Since $A^iB^i \subseteq AB$, we have $x \in (y,z)$ for some $z \in AB$. Hence

$$x \in (A^iB^i)(AB) = (A^iA)(B^iB) = A^iB^i.$$

We show next that $A^iB^i \subseteq (AB)^i$. Let $z \in A^iB^i$ and let $c \in AB$. Then $z \in xy$ for some $x \in A^i$ and $y \in B^i$, and $c \in ab$ for some $a \in A$ and $b \in B$. Hence $x \in aa'$ for some $a' \in A$, and $y \in bb'$ for some $b' \in B$. Thus

$$z \in (aa')(bb') = (ab)(a'b').$$

But $abc = ab$, since $c \in ab$ and $ab = (ab)^i$. Hence $z \in c(AB)(AB) = c(AB)$. Thus $z \in (AB)^i$, by Proposition 20(i). □

From Propositions 11 and 23 we immediately obtain

PROPOSITION 24 *If A and B are convex sets such that* $A^i \neq \emptyset$ *and* $B^i \neq \emptyset$, *then*

$$\overline{[A \cup B]} = \overline{AB}. \quad \square$$

From Proposition 23 we also obtain by induction a characterization of the intrinsic interior of a polytope:

PROPOSITION 25 *For any positive integer n and any* $a_1,...,a_n \in X$,

$$[a_1,...,a_n]^i = a_1...a_n. \quad \square$$

2 UNENDING LINEAR GEOMETRIES

A linear geometry will be said to be *unending* if it has the following property:

(U) *if* $a,b \in X$ *and* $a \neq b$, *then* $a \in [b,c]$ *for some* $c \neq a,b$.

This axiom also implies the axiom **(P)**, when all the other axioms of a linear geometry are satisfied. Moreover, as we now show, an unending linear geometry is 'generally' dense:

PROPOSITION 26 *A convex geometry is a linear geometry if it satisfies the axioms* **(L2)–(L4)** *and* **(U)**. *Furthermore, it is also dense if not all points are collinear.*

Proof By Proposition III.6 and Proposition 1, we need only show that the axiom **(D)** is satisfied if not all points of X are collinear.

Let y_1,y_2 be any two distinct points and choose $z \notin \langle y_1,y_2 \rangle$. By **(U)** we can now choose z_1 so that $z \in (y_2,z_1)$ and then x so that $z_1 \in (x,y_1)$. It follows from Proposition II.2′ that $z \in (x,y)$ for some $y \in (y_1,y_2)$. □

An unending linear geometry need not be dense if all points are collinear; for example, take $X = \mathbb{R} \setminus (0,1)$ and define the segment $[x,y]$ to be the intersection with X of the corresponding segment in $\mathbb{R}$.

It may be shown also that *a convex geometry is a linear geometry if it satisfies the axioms* **(L2)**,**(L3)**,**(P)** *and* **(U)**, i.e. these axioms imply the axiom **(L4)**. For suppose $c,d \in [a,b]$. We wish to show that $d \in [a,c] \cup [c,b]$. Evidently we may assume $a \neq b$, $c \neq a,b$ and $d \neq a,b,c$. By **(U)** we may choose e so that $b \in (a,e)$. Since $c,d \in [a,b]$, it follows from Proposition II.18(i) that $b \in [c,e] \cap [d,e]$. Since $b \neq e$, it now follows from **(L3)** that $c \in [d,e]$ or $d \in [c,e]$. If $d \in [c,e]$ then, since $b \in [d,e]$, it follows from Proposition II.18(i) that $d \in [b,c]$. Thus we now suppose $c \in [d,e]$. From $b \in [d,e]$ and $d \in [a,b]$ we obtain, by **(L2)**, $d \in [a,e]$. From $d \in [a,e]$ and $c \in [d,e]$ we obtain $d \in [a,c]$, by Proposition II.18(i) again.

Throughout the remainder of this section we assume that a set X is given on which a dense, unending linear geometry is defined.

An immediate consequence of this assumption is that, if H is a hyperplane of X, then $X \setminus H$ is *not* convex. Hence Proposition III.12 and Proposition 17 now apply to arbitrary hyperplanes. Another consequence of this assumption is that the definition of intrinsic interior may be reformulated in the following way:

PROPOSITION 27 *Suppose C is a convex set and let $c \in C$. Then $c \in C^i$ if and only if $C \cap (c,x) \neq \emptyset$ for every $x \in \langle C \rangle \setminus c$.*

Proof The necessity of the condition has already been proved in Proposition 9. Suppose now that $C \cap (c,x) \neq \emptyset$ for every $x \in \langle C \rangle \setminus c$. For any $d \in C \setminus c$, choose x so that $c \in (d,x)$. If $e \in C \cap (c,x)$, then $c \in (d,e)$ and hence $c \in C^i$. □

A convex set C will be said to be *basic* if $C^i = C$ and $\langle C \rangle = X$. For example, X itself is a basic convex set and so are the two open half-spaces associated with any hyperplane. If C and D are basic convex sets with $C \cap D \neq \emptyset$ then, by Propositions 12 and 9, $C \cap D$ is also a basic convex set. Consequently a dense unending linear geometry X can be given the structure of a *topological space* by defining a set $G \subseteq X$ to be *open* if $G = \emptyset$ or if G is a union of basic convex sets. Moreover, with this topology X is a *Hausdorff space*, by Proposition 18. (The various topological terms used here are defined in Section 4 of Chapter VIII.)

The notion of convex partition was introduced in Chapter II. Some further properties of convex partitions can be derived under the hypotheses of the present section:

PROPOSITION 28 *Let C,D be a convex partition of X. If $C^i \neq \emptyset$, then also $D^i \neq \emptyset$ and both $\overline{C},D^i$ and $\overline{D},C^i$ are convex partitions of X.*

Proof Let $x \in C^i$ and $y \in D$. If $y \in (x,z)$, then $z \in D$. We will show that actually $z \in D^i$.

Assume on the contrary that $z \notin D^i$. Then there exists $u \in D$ such that $z \in (u,v)$ implies $v \in C$. Fix any such v. Evidently $u \notin \langle x,z \rangle$. Since $x \in C^i$, there exists $w \in C$ such that $x \in (v,w)$. Then $y \in (t,w)$ for some $t \in (z,v)$. Since $w \in C$ and $y \in D$, we must have $t \in D$. Since $z \in (u,t)$ implies $t \in C$, this is a contradiction.

This proves that $D^i \neq \emptyset$. It also proves that if $y \in \overline{C} \cap D$, then $y \notin D^i$ and that if $y \in D \setminus D^i$, then $(x,y) \subseteq C$. Thus $\overline{C} \subseteq X \setminus D^i$ and $X \setminus D^i \subseteq \overline{C}$, which shows that $\overline{C},D^i$ is a convex partition of X. Moreover, C and D may now be interchanged. □

It is possible that under the hypotheses of Proposition 28 we have both $C^i = C$ and $D^i = D$. For example, take $X = \mathbb{Q}$ to be the field of rational numbers, C to be the set of all $x \in \mathbb{Q}$ with $x > 0$ and $x^2 > 2$, and $D = X \setminus C$.

For any $a,b \in X$, put

$$\begin{aligned} a/b &= \{c \in X : a \in (b,c)\} \text{ if } a \neq b, \\ &= \{a\} \text{ if } a = b. \end{aligned}$$

Thus $c \in a/b$ if and only if $a \in bc$. More generally, for any nonempty subsets A,B of X we define the *quotient* A/B by

$$A/B = \bigcup_{a \in A, b \in B} a/b.$$

In other words, A/B is the union of $A \cap B$ with the set of all points $c \in X$ such that $a \in (b,c)$ for some distinct points $a \in A$ and $b \in B$. It follows from **(U)** that $A/B \neq \emptyset$. Evidently if $A \subseteq C$ and $B \subseteq D$, then $A/B \subseteq C/D$.

The reason for the terminology and notation is that quotients possess many of the properties of ordinary fractions, provided equality is replaced by inclusion:

LEMMA 29 *If A and B are nonempty subsets of X, then*

(i) $A \subseteq B(A/B)$,
(ii) $A \subseteq (AB)/B$,
(iii) $A \subseteq B/(B/A)$.

Proof Since these properties are really just reformulations of the definitions, we only prove (i): if $a \in A$, $b \in B$ and $c \in ab$, then $c \in AB$ and $a \in (AB)/B$. □

LEMMA 30 *If A,B,C are nonempty subsets of X, then*

(i) $(A/B)/C = A/(BC) = (A/C)/B$,
(ii) $A/(B/C) \subseteq (AC)/B$,
(iii) $A(B/C) \subseteq (AB)/C$.

Proof (i) Suppose $x \in (a/b)/c$ for some $a \in A$, $b \in B$, $c \in C$. Then there exists $y \in X$ such that $y \in cx$ and $a \in by$. Hence, by **(C)″**, there exists $z \in bc$ such that $a \in xz$. Thus $x \in a/(bc)$.

On the other hand suppose $x \in a/(bc)$ for some $a \in A$, $b \in B$, $c \in C$. Then $a \in xy$ for some $y \in bc$. Hence, by **(C)″**, $a \in bz$ for some $z \in cx$. Thus $x \in (a/b)/c$.

This proves the first equality in (i) and, since $BC = CB$, the second follows.

(ii) Suppose $x \in a/(b/c)$ for some $a \in A$, $b \in B$, $c \in C$. Then there exists $y \in X$ such that $a \in xy$ and $b \in cy$. Hence, by **(P)″**, there exists $z \in ac \cap bx$. Thus $x \in (ac)/b$.

(iii) Suppose $x \in a(b/c)$ for some $a \in A$, $b \in B$, $c \in C$. Then there exists $y \in X$ such that $x \in ay$ and $b \in cy$. Hence, by **(P)″**, there exists $z \in ab \cap cx$. Thus $x \in (ab)/c$. □

LEMMA 31 *If A,B,C,D are nonempty subsets of X, then*

(i) $(A/B)(C/D) \subseteq AC/BD$,

(ii) $(A/B)/(C/D) \subseteq AD/BC$.

Proof (i) By Lemma 30(iii), $(A/B)(C/D) \subseteq ((A/B)C)/D$ and

$$(A/B)C = C(A/B) \subseteq AC/B.$$

Hence, by Lemma 30(i),

$$(A/B)(C/D) \subseteq (AC/B)/D = AC/BD.$$

(ii) By Lemma 30(ii), $(A/B)/(C/D) \subseteq ((A/B)D)/C$, and by Lemma 30(iii),

$$(A/B)D = D(A/B) \subseteq AD/B.$$

Hence, by Lemma 30(i),

$$(A/B)/(C/D) \subseteq (AD/B)/C = AD/BC. \quad \square$$

PROPOSITION 32 *If A and B are nonempty convex sets, then the quotient A/B is also a nonempty convex set.*

Proof Obviously $A/B \subseteq (A/B)(A/B)$. On the other hand, by Lemma 31(i),

$$(A/B)(A/B) \subseteq (AA)/(BB) = A/B. \quad \square$$

For any nonempty set $A \subseteq X$ we have $A \subseteq A/A$. It will now be shown that affine sets are characterized by equality in this relation:

PROPOSITION 33 *A nonempty set A is affine if and only if $A/A = A$.*

Proof Suppose first that A is affine. If $x \in A/A$, then there exist $a,a' \in A$ such that $a \in xa'$ and hence $x \in A$. Since $A \subseteq A/A$, this proves that $A = A/A$.

Suppose next that $A/A = A$ and let y,z be distinct elements of A. If $y \in (z,u)$, then $u \in y/z \subseteq A$. If $w \in (y,z)$, then $y \in (w,u)$ and $w \in y/u \subseteq A$. If $z \in (y,v)$, then $v \in z/y \subseteq A$. Thus $\langle y,z\rangle \subseteq A$, which proves that A is affine. $\square$

PROPOSITION 34 *If A is a nonempty convex set, then $\langle A\rangle = A/A$.*

Proof If $x \in A/A$, then there exist $a,a' \in A$ such that $a \in xa'$ and hence $x \in \langle A\rangle$.

On the other hand if $x \in \langle A \rangle$ then, by Proposition III.7, $x \in \langle a,a' \rangle$ for some $a,a' \in A$. Hence either $x \in A$ or $a \in (x,a')$ or $a' \in (x,a)$. In every case $x \in A/A$. □

PROPOSITION 35 *If A is a convex set and $x \in A^i$, then* $\langle A \rangle = A/x = x/A$.

Proof It is obvious that $A/x \subseteq \langle A \rangle$ and $x/A \subseteq \langle A \rangle$. Moreover $x \in A/x$ and $x \in x/A$. Suppose $y \in \langle A \rangle$ and $y \neq x$. Then, by Proposition 9, there exists a point $z \in A \cap (x,y)$. Hence $y \in z/x \subseteq A/x$. Furthermore, since $x \in A^i$, we have $x \in (z,z')$ for some $z' \in A$. Hence $x \in (z',y)$ and $y \in x/z' \subseteq x/A$. □

PROPOSITION 36 *If A is a convex set and $x \in A$, then the following statements are equivalent:*

(i) $x \in A^i$,
(ii) $A \subseteq x/A$,
(iii) x/A *is affine,*
(iv) A/x *is affine.*

Proof It follows from Proposition 35 that (i) $\Rightarrow$ (ii),(iii),(iv). To complete the proof we show that (ii),(iii),(iv) $\Rightarrow$ (i). We may assume that $|A| > 1$. Let $y \in A \setminus x$ and choose $z,w \in X$ so that $x \in (y,z)$, $y \in (x,w)$. Then $z \in x/A$ and $w \in A/x$.

If $A \subseteq x/A$, then $y \in x/A$. Thus $x \in (a,y)$ for some $a \in A$, and hence $x \in A^i$.

If x/A is affine then $y \in x/A$, since $x \in x/A$. Hence $A \subseteq x/A$ and $x \in A^i$.

If A/x is affine then $z \in A/x$, since $x \in A/x$. Thus there exists a point $a' \in A \cap (x,z)$. Then $x \in (a',y)$, and hence $x \in A^i$. □

PROPOSITION 37 *For any nonempty sets A,B,* $\langle A \cup B \rangle = \langle A/B \rangle$.

Proof Since $A \subseteq B(A/B)$, by Lemma 30(i) we have

$$A \subseteq A/A \subseteq A/(B(A/B)) = (A/B)/(A/B) \subseteq \langle A/B \rangle.$$

Similarly, since $B \subseteq A/(A/B)$, we have

$$B \subseteq B/B \subseteq (A/(A/B))/B = (A/B)/(A/B) \subseteq \langle A/B \rangle.$$

Since $A/B \subseteq \langle A \cup B \rangle$, it follows that $\langle A \cup B \rangle = \langle A/B \rangle$. □

PROPOSITION 38 *If A and B are convex sets such that $A^i \cap B^i \neq \emptyset$, then*

$$\langle A \cup B \rangle = A/B = A^i/B^i.$$

Proof If $z \in A^i \cap B^i$ then, by Propositions 35 and 20,

$$\langle A\rangle/\langle B\rangle = (A/z)/(B/z) \subseteq (zA)/(zB) = A^i/B^i \subseteq A/B \subseteq \langle A\rangle/\langle B\rangle.$$

Hence $\langle A\rangle/\langle B\rangle = A/B = A^i/B^i$. Consequently

$$\begin{aligned}(A/B)/(A/B) = (A/B)/(\langle A\rangle/\langle B\rangle) &= (A/B)/((z/A)/(z/B)) \\ &\subseteq (A/B)/((zB)/(zA)) \\ &\subseteq (A/B)/(B/A) \\ &\subseteq (AA)/(BB) = A/B.\end{aligned}$$

Hence A/B is affine, by Proposition 33, and the result now follows from Proposition 37. □

PROPOSITION 39 *If A and B are convex sets such that $A^i \neq \emptyset$ and $B^i \neq \emptyset$, then*

$$(A/B)^i = A^i/B^i \neq \emptyset.$$

Proof We show first that $(A/B)^i \subseteq A^i/B^i$. Let $x \in (A/B)^i$ and let $y \in A^i/B^i$ with $y \neq x$. Since $A^i/B^i \subseteq A/B$, we have $x \in (y,z)$ for some $z \in A/B$. Hence, by Lemma 31 and Proposition 20,

$$x \in (A^i/B^i)(A/B) \subseteq (A^iA)/(B^iB) = A^i/B^i.$$

We show next that $A^i/B^i \subseteq (A/B)^i$. If $z \in A^i/B^i$, then $z \in x/y$ for some $x \in A^i$, $y \in B^i$. Hence $x \in A^i \cap (zB^i) = A^i \cap (zB)^i$, by Proposition 23. It now follows from Proposition 38 that

$$A^i/(zB)^i = A/(zB) = \langle A \cup (zB)\rangle.$$

Thus $(A/B)/z = A/(zB)$ is affine and hence, by Proposition 36, $z \in (A/B)^i$. □

The next result may be regarded as an affine analogue of Proposition II.4, but it requires a supplementary hypothesis.

PROPOSITION 40 *If A and B are affine sets such that $A \cap B \neq \emptyset$, then*

$$\langle A \cup B\rangle = A/B = \bigcup_{a\in A, b\in B} \langle a,b\rangle.$$

Proof Since an affine set is its own intrinsic interior, and since it is obvious that

$$A/B \subseteq \bigcup_{a\in A, b\in B} \langle a,b\rangle \subseteq \langle A \cup B\rangle,$$

the result follows immediately from Proposition 38. □

The collection of all affine subsets of X, partially ordered by inclusion, is a *lattice*, since any two affine sets A and B have an infimum $A \cap B$ and a supremum $\langle A \cup B\rangle$. Our next result says that this lattice is *weakly modular*:

PROPOSITION 41 *If A,B,C are affine sets such that $C \subseteq B$ and $A \cap B \neq \emptyset$, then*

$$\langle C \cup A\rangle \cap B = \langle C \cup (A \cap B)\rangle.$$

Proof The right side is certainly contained in the left, since $C \cup (A \cap B)$ is contained in both B and $\langle C \cup A\rangle$. To show that the left side is contained in the right let $p \in A \cap B$ and $x \in \langle C \cup A\rangle \cap B$. By Proposition 40 $x \in \langle a,b\rangle$, where $a \in A$ and $b \in \langle p \cup C\rangle$. If $x \in \langle p \cup C\rangle$, then $x \in \langle C \cup (A \cap B)\rangle$. If $x \notin \langle p \cup C\rangle$ then, by Proposition III.8, $a \in \langle x \cup p \cup C\rangle$. Hence $a \in B$ and again $x \in \langle C \cup (A \cap B)\rangle$. □

Proposition 41 has the following important consequence:

PROPOSITION 42 *Suppose A_1 and A_2 are affine sets such that $A_1 \cap A_2 \neq \emptyset$. Then $A_1 \cap A_2$ has finite codimension in A_2 if and only if A_1 has finite codimension in $\langle A_1 \cup A_2\rangle$, and the two codimensions are then equal.*

In particular, if A_1 and A_2 are finite-dimensional, then $\langle A_1 \cup A_2\rangle$ is also finite-dimensional and

$$\dim A_1 + \dim A_2 = \dim (A_1 \cap A_2) + \dim \langle A_1 \cup A_2\rangle.$$

Proof By Proposition I.22, we need only show that if A_1 has finite codimension m in $\langle A_1 \cup A_2\rangle$, then $A_1 \cap A_2$ has finite codimension n in A_2, where $n \leq m$. Let B be a basis for $A_1 \cap A_2$. Then A_1 has a basis of the form $B \cup B_1$ and $\langle A_1 \cup A_2\rangle$ has a basis of the form $B \cup B_1 \cup B_2'$, where $B_2' \subseteq A_2$ and $|B_2'| = m$. By Proposition 41, with A,B,C replaced by $A_1,A_2,\langle B_2'\rangle$, we have

$$A_2 = <A_1 \cup B_2'> \cap A_2 = <B_2' \cup B>.$$

Hence A_2 has a basis of the form $B \cup B_2''$, where $B_2'' \subseteq B_2'$. □

The hypothesis that X is unending cannot be omitted in Propositions 40–42. For example, let X be the convex subset of $\mathbb{R}^3$ consisting of the origin (0,0,0) and all points (x,y,z) with $x,y,z > 0$. If we take A to be the plane consisting of all points $(x,y,z) \in X$ with $y = x$ and B to be the plane consisting of all points $(x,y,z) \in X$ with $y = 2x$, then $A \cap B = \{(0,0,0)\}$ and hence

$$\dim (A \cap B) + \dim <A \cup B> < \dim A + \dim B.$$

Furthermore, if we take C to be the line consisting of all points $(x,y,z) \in B$ with $z = x$, then $<A \cup C> = X$ and $A \cap B \subseteq C$, but the point $(1,2,2) \in B$ does not lie on any line $<a,c>$ with $a \in A$ and $c \in C$.

As an application of Proposition 42 we prove

PROPOSITION 43 *Let H_+,H_- be the open half-spaces of X associated with a hyperplane H and let A be an affine subset of X. If A contains points of both H_+ and H_-, then $L: = A \cap H$ is a hyperplane of A and $A \cap H_+$, $A \cap H_-$ are the open half-spaces of A associated with this hyperplane.*

Proof The hypotheses imply that $L \neq A, \emptyset$. Hence L is a hyperplane of A, by Proposition 42. Since $A \cap H_+$ is disjoint from L, it is contained in an open half-space L_+ of A associated with the hyperplane L. Let L_- be the other open half-space of A associated with the hyperplane L. If $x \in A \cap H_+$ and $y \in A \cap H_-$, then (x,y) contains a point $z \in L$ and hence $y \in L_-$. Since $A \cap H_+ \subseteq L_+$, $A \cap H_- \subseteq L_-$ and A is the union of L, $A \cap H_+$ and $A \cap H_-$, we must actually have $A \cap H_+ = L_+$, $A \cap H_- = L_-$. □

3 CONES

In this section we assume, initially, that a set X is given on which a linear geometry is defined.

For any $z,x \in X$ with $z \neq x$, we define the *ray* $[z,x>$ through x from z to be the set of all points y such that $y \in [z,x]$ or $x \in (z,y)$. We will sometimes find it convenient to denote by $(z,x>$ the 'open' ray $[z,x> \setminus z$. Evidently if $x' \in (z,x>$, then $[z,x'> = [z,x>$ and $(z,x'> = (z,x>$.

A set $K \subseteq X$ is said to be a *cone* with *vertex* z if $x \in K$ implies $[z,x> \subseteq K$. The vertex need not be uniquely determined; the set of all vertices of a cone will be called its *vertex set*.

For example, the empty set is a cone with X as its vertex set, and the singleton $\{z\}$ is a cone with vertex z. An arbitrary union of rays from z is a cone with vertex z. A nonempty set L is affine if and only if it is a cone with vertex set L.

The definition of a cone does not require it to be convex, but the convex case is of greater interest. Thus if L is a nonempty affine set and $x \in X \setminus L$ then, by Proposition III.12, the closed half-space of $<x \cup L>$, which is associated with the hyperplane L and which contains x, is a convex cone with every point of L as a vertex.

It follows at once from the definition that if z is a vertex of a nonempty cone K, then $z \in K$. Moreover, if z and z' are vertices of K, then $<z,z'> \subseteq K$. In fact, as we now show, every point of $<z,z'>$ is a vertex.

PROPOSITION 44 *The vertex set of a cone is an affine set.*

Proof Let K be a cone and let u,v be vertices of K. We wish to show that if $w \in <u,v>$ and $x \in K$, then $y \in [w,x]$ or $x \in (w,y)$ implies $y \in K$.

Suppose first that $w \in [u,v]$. If $y \in [w,x]$, then there exists a point $y' \in [v,x]$ such that $y \in [u,y']$. Moreover $y' \in K$, since v is a vertex, and hence $y \in K$, since u is a vertex. On the other hand if $x \in (w,y)$, then there exists a point $x' \in [v,y]$ such that $x \in [u,x']$. Moreover $x' \in K$, since u is a vertex, and hence $y \in K$, since v is a vertex.

Suppose next that $u \in [v,w]$. If $y \in [w,x]$, then there exists a point $y' \in [u,x] \cap [v,y]$. Moreover $y' \in K$, since u is a vertex, and hence $y \in K$, since v is a vertex. Similarly if $x \in (w,y)$, then there exists a point $x' \in [v,x] \cap [u,y]$. Moreover $x' \in K$, since v is a vertex, and hence $y \in K$, since u is a vertex.

Since u and v can be interchanged, this completes the proof. □

For convex cones we can say more:

PROPOSITION 45 *The vertex set of a convex cone K is a face of K.*

Proof Let z be a vertex of K and suppose $z \in (z',z'')$, where $z',z'' \in K$. Then $\langle z',z''\rangle \subseteq K$, since z is a vertex. Let $x \in K \setminus \langle z',z''\rangle$. If $y \in [z',x]$ then $y \in K$, since K is convex. On the other hand if $x \in (z',y)$, there exists a point $x' \in (z,y) \cap (z'',x)$. Then $x' \in K$, since K is convex, and hence $y \in K$, since z is a vertex. Consequently z' is a vertex, and z'' likewise. The result now follows from Proposition II.6. □

It follows that if K is a nonempty convex cone and L the affine set of all vertices of K, then $K \setminus L$ is convex.

Let $\mathcal{K}_L$ denote the collection of all convex cones whose vertex set contains L, where L is a nonempty affine set. It is easily seen that $\mathcal{K}_L$ is a normed alignment on X. We define the *L-conic hull* $[L,S\rangle$ of any set $S \subseteq X$ to be the intersection of all convex cones in $\mathcal{K}_L$ which contain S.

Evidently $[L,S\rangle \in \mathcal{K}_L$. Furthermore our notations are consistent, since for $L = \{z\}$ and $S = \{x\}$ the L-conic hull of S is the ray $[z,x\rangle$. From the definition of L-conic hull we immediately obtain the following properties:

$$[S] \subseteq [L,S\rangle \subseteq \langle S \cup L\rangle,$$
$$[L,S\rangle = [L,[S]\rangle,$$
$$[L',S\rangle \subseteq [L,S\rangle,$$

where L' is any nonempty affine subset of L.

When L is a singleton, L-conic hulls have a very simple structure:

PROPOSITION 46 *If C is a convex set and $z \in X$, then*

$$[z,C\rangle = \bigcup_{x \in C} [z,x\rangle.$$

Proof Obviously the right side is contained in the left. On the other hand, the right side contains C and is a cone with vertex z. Thus it merely remains to show that it is convex.

Suppose $y' \in [z,x'\rangle$ and $y'' \in [z,x''\rangle$, where $x',x'' \in C$. We need only show that if $y \in (y',y'')$, then $y \in [z,x\rangle$ for some $x \in [x',x'']$. If $y' \in [x',z]$ and $y'' \in [x'',z]$, then $y \in [x,z]$ for some $x \in [x',x'']$. If $x' \in (y',z)$ and $y'' \in [x'',z]$, then $y \in [z,z']$ for some $z' \in [x'',y']$ and there exists a point $x \in [x',x''] \cap [z,z']$. Hence either $y \in [x,z]$ or $x \in (y,z)$. Finally, if $x' \in (y',z)$ and $x'' \in (y'',z)$, then there exist a point $z' \in [y,z] \cap [x',y'']$ and also a point $x \in [x',x''] \cap [z,z']$. Hence $x \in (y,z)$. □

In order to progress further *we assume, throughout the remainder of this section, that the linear geometry X is both dense and unending.*

PROPOSITION 47 *Let L be a nonempty affine set. If K is a cone whose vertex set contains L, then*

$$K = LK = K/L = LK/L.$$

Conversely, if $K = LK/L$ and $L \subseteq K$, then K is a cone whose vertex set contains L.

Proof It follows directly from the definition of a cone and **(U)** that, if K is a cone whose vertex set contains L, then $K = LK$ and $K = K/L$. These two relations obviously imply $K = LK/L$. Suppose on the other hand that $K = LK/L$ and $L \subseteq K$. Let $z \in L$ and $x \in K \setminus z$, so that $x \in y'/z'$ for some $y' \in LK$ and $z' \in L$. We wish to show that $[z,x> \subseteq K$.

If $x \in (y,z)$, then $y' \in xz' \subseteq yzz'$ and hence $y' \in yz''$ for some $z'' \in zz'$. Thus $z'' \in L$, and $y \in LK/L = K$. If $y \in (x,z)$ then, by **(P)**$''$, there exists a point $w \in yz' \cap y'z$. Thus $w \in (LK)L = LK$ and $y \in LK/L = K$. Furthermore $z \in K$, by hypothesis. □

PROPOSITION 48 *For any nonempty convex set C and any nonempty affine set L,*

$$[L,C> = [L \cup C]/L = LC/L \cup L.$$

Proof Put $K = [L \cup C]/L$. Then K is a convex set containing L. From $C \subseteq L(C/L) \subseteq LC/L$ we obtain $C \subseteq K$ and

$$C/L \subseteq (LC/L)/L = LC/L.$$

Since $[L \cup C] = LC \cup L \cup C$, by Proposition 21, it follows that $K = LC/L \cup L$. Hence

$$LK/L = (L(LC/L))/L \cup L \subseteq (LC/L)/L \cup L = K.$$

On the other hand, since $L \subseteq LK/L$ and $LC/L \subseteq LK/L$, we also have $K \subseteq LK/L$. Consequently $K = LK/L$ and K is a cone whose vertex set contains L, by Proposition 47. Furthermore, if K' is any convex cone containing C whose vertex set contains L, then

$$K = LC/L \cup L \subseteq LK'/L = K'.$$

Consequently $K = [L,C\rangle$. □

We consider next the structure of the intrinsic interior and convex closure of a convex cone. If some vertex of a convex cone K is contained in its intrinsic interior K^i then, by Proposition 45, every point of K is a vertex. Hence K is affine and $K = K^i = \overline{K}$. We now treat the general case:

PROPOSITION 49 *If K is a convex cone whose vertex set contains the nonempty affine set L, then $K^i \cup L$ and $\overline{K}$ are also convex cones whose vertex sets contain L.*

Proof Obviously we may assume that $K^i \neq \emptyset$. Then $H := K^i \cup L$ is convex, since K^i and L are convex and $L \subseteq K$. To show that H is a cone whose vertex set contains L, we need only show that if $z \in L$, $x \in K^i \backslash L$ and $x \in (y,z)$, then $y \in K^i$. We certainly have $y \in K$. Let $y' \in K \backslash y$ and choose $y'' \in X$ so that $y \in y'y''$. Then $x \in y'y''z$ and hence $x \in y'z'$ for some $z' \in y''z$. Since $x \in K^i$, it follows that $x \in y'x'$ for some $x' \in xz' \cap K$. Since $x \in yz$ and $z' \in y''z$, it follows that $x' \in x''z$ for some $x'' \in yy''$. Moreover $x'' \in K$, since $x' \in K$ and z is a vertex of K. Since $y \in y'x''$, it follows that $y \in K^i$.

To prove that $\overline{K}$ is also a convex cone whose vertex set contains L, we need only show that $\overline{K}$ contains the ray $[z,y\rangle$ for any $y \in \overline{K} \backslash K$ and $z \in L$. Choose any $y' \in [z,y\rangle$, $x \in K^i$ and $x' \in xy'$. Suppose first that $y' \in (z,y)$. Then, by **(C)″**, $x' \in zz'$ for some $z' \in xy$. Consequently $z' \in K^i$ and hence $x' \in K^i$. This proves that $y' \in \overline{K}$. Suppose next that $y \in (z,y')$. Then, by **(P)″**, there exists a point $z'' \in zx' \cap xy$. Consequently $z'' \in K^i$ and hence $x' \in K^i$, by the first part of the proof. Thus again $y' \in \overline{K}$. □

PROPOSITION 50 *Let L be a nonempty affine set and K a convex cone whose vertex set is L. Then $K^\dagger = L/K$ is also a convex cone whose vertex set is L. Moreover $K \cap K^\dagger = L$, $(K^\dagger)^\dagger = K$, and $K^\dagger = z/K$ for any $z \in L$.*

Proof We may assume $K \neq L$. Let $z \in L$ and put $K^z = z/K$. Thus K^z is the union of $\{z\}$ with the set of all points $x \in X$ such that $z \in (x,y)$ for some $y \in K$. Evidently K^z is a cone with vertex z, and $K^z \neq L$ since $K \neq L$. Moreover K^z is convex, by Proposition II.19″, and $L \subseteq K^z$ since $L \subseteq K$.

In fact $L = K \cap K^z$. For if $z' \in K \cap K^z$ and $z' \neq z$, then $z \in (z',z'')$ for some $z'' \in K$ and hence $z' \in L$, since L is a face of K by Proposition 45. Since it is clear from the definition that $(K^z)^z = K$, it follows that the vertex set of K^z is $K^z \cap K = L$.

It only remains to show that $K^z = L/K$. In fact, since it is obvious that $K^z \subseteq L/K$, we need only show that if $x' \in L/K$, then $x' \in K^z$. There exist $y' \in K$ and $z' \in L$ such that $z' \in x'y'$. If we choose x'' so that $z \in x'x''$, then there exists a point $y \in zy' \cap z'x''$ by **(P)″**. It follows that $y \in K$, since K is convex, and hence that $x'' \in K$, since $z' \in L$. Consequently $x' \in K^z$. □

We will call $K^\dagger$ the convex cone *symmetric* to the convex cone K.

It is easily seen that, for any hyperplane H, the two closed half-spaces associated with H are symmetric convex cones with H as their common vertex set. Furthermore, if H_+ and H_- are the two open half-spaces associated with H then, for any nonempty affine set $L \subseteq H$, $H_+ \cup L$ and $H_- \cup L$ are symmetric convex cones with L as their common vertex set. This provides an example of a convex cone with a different vertex set from that of its convex closure.

PROPOSITION 51 *Let L be a nonempty affine set and let $K,K^\dagger$ be symmetric convex cones with the common vertex set L. Then*

(i) $K^i \neq \varnothing$ *if and only if* $(K^\dagger)^i \neq \varnothing$,

(ii) $K^i \cup L$ *and* $(K^\dagger)^i \cup L$ *are symmetric convex cones with the common vertex set* L,

(iii) $\overline{K}$ *and* $\overline{K^\dagger}$ *are also symmetric convex cones.*

Proof We show first that the vertex set of $K^i \cup L$ is exactly L. By Proposition 49 the vertex set of $K^i \cup L$ contains L. Assume that $K^i \cup L$ has a vertex $w \in K^i \setminus L$. Then for some $x \in K$ the ray $[w,x\rangle$ is not wholly contained in K. However, the line $\langle x,w\rangle$ contains points $w',w'' \in K^i$ such that $w \in (w',w'')$. Moreover w',w'' are vertices of $K^i \cup L$, since the set of all vertices is a face of $K^i \cup L$. Hence $\langle x,w\rangle = \langle w',w''\rangle \subseteq K^i \cup L$, which is a contradiction.

If $K^i \neq \varnothing$ then, by Proposition 39, $(K^\dagger)^i = L/K^i \neq \varnothing$. Since the relation between K and $K^\dagger$ is symmetric, this proves (i). Furthermore, since $L/L = L$,

$$(K^\dagger)^i \cup L \;=\; L/K^i \cup L \;=\; L/(K^i \cup L) \;=\; (K^i \cup L)^\dagger.$$

This completes the proof of (ii).

To prove (iii) we may assume $K^i \neq \emptyset$. Fix $z \in L$. Then, by Proposition 39, we also have $(K^\dagger)^i = z/K^i$. Suppose $y' \in \overline{K}^\dagger$ and $x' \in (K^\dagger)^i$ with $x' \neq y'$. Then $z \in yy'$ for some $y \in \overline{K}$ and $z \in xx'$ for some $x' \in K^i$. If $w' \in x'y'$ then $z \in ww'$ for some $w \in xy$, by Proposition II.19″. Thus $w \in K^i$ and hence $w' \in (K^\dagger)^i$. This proves that $y' \in \overline{K^\dagger}$.

Thus $\overline{K}^\dagger \subseteq \overline{K^\dagger}$. Replacing K by $K^\dagger$, we obtain $\overline{K^\dagger}^\dagger \subseteq \overline{K}$. Since $K_1 \subseteq K_2$ implies $z/K_1 \subseteq z/K_2$, it follows that also $\overline{K^\dagger} \subseteq \overline{K}^\dagger$. □

4 POLYHEDRA

Throughout this section we will assume that a set X is given on which a dense unending linear geometry is defined. A set $P \subseteq X$ is said to be a *polyhedron* if it is affine or if it is the intersection of finitely many closed half-spaces of its affine hull $<P>$.

Examples of polyhedra in $\mathbb{R}^2$ are a strip bounded by two non-intersecting lines and a sector bounded by two intersecting lines.

Obviously any polyhedron is convex and, by Proposition IV.19, any polytope is a polyhedron. It is easily seen that a one-dimensional polyhedron is either a line $<a,b>$ or a ray $[a,b>$ or a segment $[a,b]$.

If P is an affine set then, by Proposition II.6, its only faces are P and $\emptyset$. We are going to show that the facial structure of polyhedra which are not affine resembles that of polytopes.

If P is a polyhedron which is not affine, then there exist hyperplanes $H_1,\dots,H_m$ of $<P>$ with associated open half-spaces H_j^+, H_j^- ($j = 1,\dots,m$) such that

$$(*) \qquad P = \bigcap_{j=1}^m \overline{H_j^+}.$$

The representation (*) will be said to be *redundant* if some term $\overline{H_j^+}$ can be omitted and *irredundant* otherwise. Evidently if the representation (*) is redundant, then an irredundant representation can be obtained from it by omitting certain terms.

PROPOSITION 52 *If a nonempty polyhedron P has the irredundant representation (*), then* $\overline{P} = P$ *and* $P^i = \bigcap_{j=1}^m H_j^+ \neq \emptyset$. *Furthermore,*

(i) *the distinct facets of P are* $F_j = P \cap H_j$ $(j = 1,...,m)$,
(ii) *any nonempty proper face of P is a polyhedron, is contained in a facet of P, and is the intersection of those facets of P which contain it,*
(iii) *a face of P is affine if and only if no nonempty face of P is properly contained in it.*

Proof Put

$$M_k = \bigcap_{j \neq k} \overline{H_j^+}.$$

Then $P = M_k \cap \overline{H_k^+}$ and $F_k = P \cap H_k = M_k \cap H_k$ for each $k \in \{1,...,m\}$. Since P is not contained in H_k, M_k is not contained in $\overline{H_k^-}$. On the other hand, since the representation (*) is irredundant, M_k is not contained in $\overline{H_k^+}$. Thus M_k contains a point of H_k^+ and a point of H_k^-, and hence also a point of H_k.

Thus $F_k \neq \emptyset$. We are going to show that actually $\langle F_k \rangle = H_k$. Assume on the contrary that there exists a point $x \in H_k$ such that $x \notin \langle F_k \rangle$. By Proposition III.7 we have $x \in \langle x',x'' \rangle$ for some $x',x'' \in P$. Then $x \neq x',x''$ and $x',x'' \notin H_k$. Thus $x',x'' \in M_k \cap H_k^+$. Since $x \notin [x',x'']$, we may choose the notation so that $x'' \in (x,x')$. Pick $y \in M_k \cap H_k^-$. Then (x',y) contains a point $y' \in M_k \cap H_k$. Since $y' \neq x$, the points x,x',y are not collinear and there exists a point $y'' \in (x'',y) \cap (x,y')$. Then $y'' \in M_k \cap H_k$. Since $x \in \langle y',y'' \rangle$ and $y',y'' \in F_k$, this is a contradiction.

Thus $P \cap \langle F_k \rangle = F_k$. Since also $P \setminus F_k = M_k \cap H_k^+$, it follows from Proposition IV.1 that F_k is a face of P. In fact F_k is a facet of P, since H_k is a hyperplane of $\langle P \rangle$. Moreover F_k is itself a polyhedron, since

$$F_k = \bigcap_{j \neq k} \overline{H_j^+} \cap H_k$$

and either $H_k \subseteq \overline{H_j^+}$ or $\overline{H_j^+} \cap H_k$ is a closed half-space of $H_k = \langle F_k \rangle$, by Proposition 43.

We now show that any proper face F of P is contained in some facet F_k. It is enough to show that $F \subseteq H_k$ and in fact, by Proposition 3, it is enough to show that $F \subseteq \bigcup_{j=1}^m H_j$. Assume on the contrary that there exists a point $x \in F$ such that $x \in \bigcap_{j=1}^m H_j^+$. If $y \in P \setminus x$ then $x \in (y,z)$ for some $z \in \langle P \rangle$, by (U). Moreover, since $y \in \bigcap_{j=1}^{m} \overline{H_j^+}$, we can choose $z \in \bigcap_{j=1}^{m} H_j^+$, by (**D**). Then $z \in P$. Since F is a face of P, it follows that $y \in F$. Thus $F = P$, which is a contradiction.

This shows, in particular, that $F_1,...,F_m$ are the only facets of P. Since the representation (*) is irredundant, they are certainly distinct.

It is obvious that $\bigcap_{j=1}^{m} H_j^+ \subseteq P^i$. Since no element of P^i is contained in a facet F_j, we must actually have $P^i = \bigcap_{j=1}^{m} H_j^+$. Moreover $P^i \neq \emptyset$ since, by Proposition 3, P is not contained in $\bigcup_{j=1}^{m} H_j$. If $y \in \overline{P}$ and $x \in P^i$, then $(x,y) \subseteq H_j^+$ $(j = 1,...,m)$ and hence $y \in P$. Thus $\overline{P} = P$.

Let F be any nonempty proper face of P. As we have seen, $F \subseteq F_k$ for some k. If $F = F_k$, then F is a polyhedron. If $F \neq F_k$, then F_k is not affine. By what we have already proved, the facets of F_k have the form

$$\bigcap_{j \neq k,i} \overline{H_j^+} \cap H_k \cap H_i = F_k \cap F_i$$

for some $i \neq k$. Since F is a proper face of F_k, it follows that $F \subseteq F_k \cap F_i$ for some $i \neq k$ such that $F_k \cap F_i$ is a facet of F_k. Moreover $F = F_k \cap F_i$ if and only if F is a facet of F_k and in this case F is a polyhedron. If $F \neq F_k \cap F_i$, then $F_k \cap F_i$ is not affine and the argument can be repeated. Since the process must eventually terminate, this yields (ii) and (iii). □

COROLLARY 53 *A polyhedron has only finitely many faces, and a nonempty polyhedron has at least one nonempty affine face.* □

COROLLARY 54 *If F and G are faces of a polyhedron P such that $F \subset G$, then there exists a finite sequence $F_0,F_1,...,F_r$ of faces of P, with $F_0 = F$ and $F_r = G$, such that F_{i-1} is a facet of F_i $(i = 1,...,r)$.* □

It may be noted that, in Corollary 54, the length of the sequence is the codimension of $\langle F \rangle$ in $\langle G \rangle$ and consequently does not depend on the choice of sequence.

COROLLARY 55 *If F_1 is a facet of a polyhedron P and F a nonempty facet of F_1, then F is contained in exactly one other facet F_2 of P and $F = F_1 \cap F_2$.*

Proof This follows at once from Proposition 52 and Proposition IV.7. □

PROPOSITION 56 *If F,G,H are faces of a polyhedron P such that $F \subseteq G \subseteq H$, then there exists a face G' of P such that $F \subseteq G' \subseteq H$, $F = G \cap G'$, and every face of P which contains both G and G' also contains H.*

Proof The proof of the corresponding Proposition IV.18 for polytopes remains valid if we replace differences of dimension by codimension. □

PROPOSITION 57 *If P is a polyhedron and A an affine set, then $P \cap A$ is a polyhedron.*

Proof Since $P \cap A = P \cap A \cap \langle P \cap A \rangle$, we may assume that $A = \langle P \cap A \rangle$. Evidently we may assume also that $A \neq \langle P \rangle$, that $P \cap A \neq \emptyset, A$ and that P is not affine. Let P have the irredundant representation (*). Then A is not contained in $\overline{H_j^+}$ for at least one j, and we may choose the notation so that $A \subseteq \overline{H_j^+}$ if and only if $j > n$. Then

$$P \cap A = \bigcap_{j=1}^{n} \overline{H_j^+} \cap A$$

and, by Proposition 43, $\overline{H_j^+} \cap A$ is a closed half-space of A associated with the hyperplane $H_j \cap A$ for each $j \in \{1,...,n\}$. Since $A = \langle P \cap A \rangle$, it follows that $P \cap A$ is a polyhedron. □

PROPOSITION 58 *The intersection of two polyhedra is again a polyhedron.*

Proof By Proposition 57 it is sufficient to show that if P is a polyhedron which is not affine, and if $\overline{L^+}$ is a closed half-space of an affine set A associated with a hyperplane L of A, then $P \cap \overline{L^+}$ is a polyhedron. Obviously we may assume that $A \neq \langle P \rangle$ and $P \cap \overline{L^+} \neq \emptyset, P$. Since $P \cap A$ is a polyhedron and $P \cap \overline{L^+} = P \cap A \cap \overline{L^+}$, we may further assume that $P \subseteq A$. Then $H = \langle P \rangle \cap L$ is a hyperplane of $\langle P \rangle$ and $\overline{H^+} = \langle P \rangle \cap \overline{L^+}$ is a closed half-space of $\langle P \rangle$ associated with this hyperplane. Hence $P \cap \overline{L^+} = P \cap \overline{H^+}$ is a polyhedron. □

Finally we will determine under what conditions a polyhedron is a polytope or a cone.

LEMMA 59 *If P is a polyhedron, $x \in P^i$ and $y \in \langle P \rangle \setminus P$, then (x,y) contains a point of $P \setminus P^i$.*

Proof Since P is not affine, it has an irredundant representation (*). Then $y \in H_k^-$ for some k. Since $x \in H_k^+$, there exists a point $y'' \in (x,y) \cap H_k$. If $y'' \notin P$ this argument can be repeated with y'' in place of y. Since the process must terminate after at most m steps, (x,y) contains a point $y' \in P \cap H_j$ for some j. Since y' belongs to a facet of P, $y' \notin P^i$. □

PROPOSITION 60 *A polyhedron containing more than one point is a polytope if and only if it does not contain an entire ray of its affine hull.*

Proof We need only show that if P is a polyhedron containing more than one point, but not containing any entire ray, then P is a polytope. Evidently P is not affine nor a closed half-space of its affine hull. Since any minimal nonempty face of P is affine, it must be a singleton and hence an extreme point of P. It follows that P has finite dimension $d \geq 1$. If $d = 1$, then P is a segment and hence a polytope. We assume that $d > 1$ and the result holds for polyhedra of dimension less than d. Then any proper face of P is a polytope.

Let $x \in P^i$ and $y \in \langle P \rangle \backslash P$. By Lemma 59, (x,y) contains a point $y' \in P \backslash P^i$. Since P does not contain a ray there exists also a point $z \in \langle P \rangle \backslash P$ such that $x \in (y,z)$, and in the same way (x,z) contains a point $z' \in P \backslash P^i$. Since $x \in (y',z')$ and y',z' are contained in one or other of the finitely many proper faces of P, it follows from the induction hypothesis that P is a polytope. □

PROPOSITION 61 *A nonempty polyhedron P is a cone if and only if it has a unique nonempty affine face V. Moreover V is then the vertex set of P.*

Proof Let P be a nonempty polyhedron which is not affine. By Proposition 52, the nonempty affine faces of P are the minimal nonempty faces.

Suppose first that P is a cone. Then the set V of all vertices of P is an affine face of P, by Propositions 44 and 45. If $v \in V$ and $x \in P \backslash V$, then $[v,x\rangle \subseteq P$ and hence $x \in (v,x')$ for some $x' \in P$. Consequently any face of P containing x must contain V and thus cannot be affine.

Suppose next that P has a unique nonempty affine face V. Let d be the codimension of $\langle V \rangle$ in $\langle P \rangle$. If $d = 1$, then P is a closed half-space associated with the hyperplane $\langle V \rangle$ of $\langle P \rangle$. Thus P is a cone with vertex set V. We now suppose $d > 1$ and use induction on d. Assume that, for some $v \in V$ and some $x \in P \backslash v$, the ray $[v,x\rangle$ is not entirely contained in P. Then $x \notin V$ and $x \in (v,y)$ for some $y \in \langle P \rangle \backslash P$. Hence, by Lemma 59, $[x,y)$ contains a point $x' \in P \backslash P^i$. Thus x' belongs to a proper face F of P. Hence $V \subseteq F$ and $x \in F$. But then, by the induction hypothesis, F is a cone and $[v,x\rangle \subseteq F$, which is a contradiction. □

5 NOTES

The concepts of intrinsic interior and convex closure, for convex sets in a real vector space, are discussed (among other things) by Bair and Fourneau (1975/80). However, the definition of convex closure used here is slightly different from theirs and has the advantage that in Proposition 7, for example, it need not be assumed that $C^i \neq \emptyset$.

The properties of products and quotients in 'join spaces' are extensively treated by Prenowitz (1961) and Prenowitz and Jantosciak (1979). They not only replace geometry by algebra, but also obviate special consideration of degenerate cases.

Proposition 26 is already contained in Veblen (1904). Proposition 41 plays a role in the lattice-theoretic characterization of affine geometry, which was initiated by Menger (1936); see, for example, Jónsson (1959), where 'weakly modular' is called 'special', and Maeda and Maeda (1970). Proposition 42, for the finite-dimensional case, is already contained in Grassmann (1844), §126. Cones are discussed for more general situations than Euclidean space by Prenowitz and Jantosciak (1979) and Rubinštein (1964).

The idea of basing the theory of polyhedra on Proposition 3 is taken from Grünbaum (1967). The usual analytic treatment replaces the theory of polyhedra by the theory of linear inequalities; see, for example, Stoer and Witzgall (1970), Tschernikow (1971) and Schrijver (1989). It should be mentioned, however, that two significant results of this theory do not carry over to the more general situation considered here. This is illustrated by the following simple example.

Let X be the open unit disc of $\mathbb{R}^2$, with the usual definition of segments, and let P be the set of all points of X in the (closed) first quadrant. Then P is a polyhedron with two facets. However, P is not the convex hull of its facets, although it is neither affine nor a closed half-space. Furthermore, P is not the convex hull of finitely many points and rays, although it is a polyhedral cone with the origin as vertex.

VI
Desargues

There are many examples of the dense unending linear geometries studied in the previous chapter, besides vector spaces over an ordered division ring. We give several such examples here, including the projective plane over the field of rational numbers. This example is not atypical, since in the next chapter we will show that any dense linear geometry of dimension greater than 2 is isomorphic to a subset of a *projective space* over an ordered division ring. In the present chapter we lay the groundwork for this result by giving a brief introduction to projective geometry and by showing that any dense linear geometry of dimension greater than 2 has the *Desargues property*. Many dense two-dimensional linear geometries do not have the Desargues property, but we establish several properties of those which do. We show also that a factor geometry of any dense linear geometry is a dense linear geometry with the Desargues property.

1 INTRODUCTION

The results which have already been established show that dense, unending linear geometries closely resemble vector spaces over an ordered division ring, which are indeed a special case. Nevertheless, there are several other examples:

EXAMPLE 1 Let X be any dense totally ordered set which has no least or greatest element. For any $x,y \in X$ define the segment $[x,y]$ to be the set of all $z \in X$ such that $x \leq z \leq y$ or $y \leq z \leq x$, according as $x \leq y$ or $y \leq x$.

For example, we may take X to be the set of all ordered pairs (ξ_1,ξ_2) of real numbers with the total ordering defined by $(\xi_1,\xi_2) \prec (\eta_1,\eta_2)$ if $\xi_1 < \eta_1$, or if $\xi_1 = \eta_1$ and $\xi_2 < \eta_2$.

EXAMPLE 2 Let X be the open hemisphere consisting of all points $x = (\xi_0, \xi_1, \ldots, \xi_n) \in \mathbb{R}^{n+1}$ with $\xi_0^2 + \xi_1^2 + \ldots + \xi_n^2 = 1$ and $\xi_0 > 0$. For any $x, y \in X$ define the segment $[x,y]$ to be $\{x\}$ if $x = y$ and otherwise to be the arc joining x and y in X of the circle through x and y with centre at the origin.

It is easily seen that the *spherical geometry* thus defined is a dense unending linear geometry. Indeed the only axioms which are not obviously satisfied are **(C)** and **(P)**. But a segment in X is the projection from the origin onto the unit sphere of the usual segment in the enveloping space $\mathbb{R}^{n+1}$, and so **(C)** follows from the corresponding property in $\mathbb{R}^{n+1}$. Since X is obviously dense, **(P)** follows from Proposition V.1.

EXAMPLE 3 Let X be the 'upper' half-space consisting of all points $x = (\xi_0, \xi_1, \ldots, \xi_n) \in \mathbb{R}^{n+1}$ with $\xi_0 > 0$, and let H be the bounding hyperplane consisting of all points $x = (\xi_0, \xi_1, \ldots, \xi_n)$ with $\xi_0 = 0$. For any $x, y \in X$ with $x \neq y$ define the segment $[x,y]$ to be the arc joining x and y in X of the circle through x and y which is orthogonal to H (or the segment joining x and y of the line through them, if this line is orthogonal to H).

We omit a detailed proof that the *hyperbolic geometry* thus defined is a dense unending linear geometry.

EXAMPLE 4 Another example is the *refracted plane*, constructed by Moulton (1902). Let $X = \mathbb{R}^2$ and define the segment $[x,y]$ to be the ordinary segment joining x and y, except in the case where the slope of the latter segment is positive and the

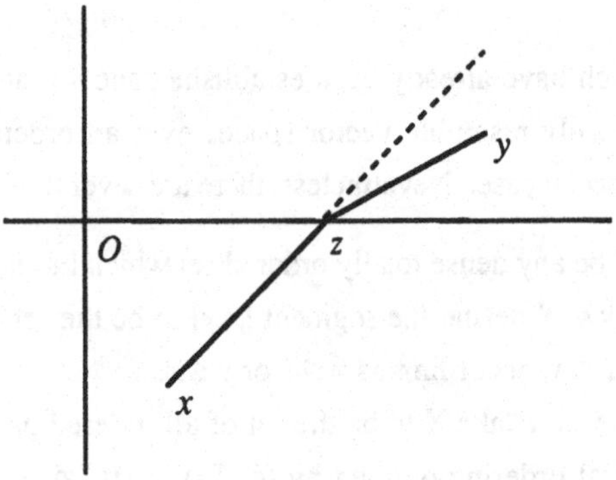

Figure 1: Moulton's refracted plane

second coordinates of x and y have opposite signs. In the excepted case, if $x = (\xi_1,\xi_2)$ and $y = (\eta_1,\eta_2)$, where $\xi_1 < \eta_1$ and $\xi_2 < 0 < \eta_2$, we define the segment $[x,y] = [y,x]$ to be the union of the segments $[x,z]$ and $[z,y]$, where $z = (\zeta_1,0)$ with $\zeta_1 = (\eta_2\xi_1 - \mu\xi_2\eta_1)/(\eta_2 - \mu\xi_2)$, for some fixed real number $\mu \in (0,1)$. In other words, z is chosen on the 'horizontal' axis so that the slope of the line <*z,y*> is μ times the slope of the line <*x,z*>, as shown in Figure 1. It is readily verified that, with this definition, X is a dense unending linear geometry.

EXAMPLE 5 The preceding example is a special case of a vastly more general construction due to Busemann (1955). Define a metric on $\mathbb{R}^2$ by putting (say)

$$d(x,y) = \max\{|\xi_1 - \eta_1|, |\xi_2 - \eta_2|\} \text{ if } x = (\xi_1,\xi_2),\ y = (\eta_1,\eta_2).$$

Let $\mathcal{G}$ be a family of subsets of $\mathbb{R}^2$ such that

(i) for each subset $g \in \mathcal{G}$, there is a continuous injective map $\tau \to g(\tau)$ of $\mathbb{R}$ onto g, so that $g = \{g(\tau): \tau \in \mathbb{R}\}$, and $d(g(\tau),g(0)) \to \infty$ as $\tau \to \pm\infty$;
(ii) for any two distinct points $a,b \in \mathbb{R}^2$, there is exactly one subset $g_{ab} \in \mathcal{G}$ which contains both a and b.

The hypothesis (i) says that each $g \in \mathcal{G}$ is homeomorphic to $\mathbb{R}$ and is a closed subset of $\mathbb{R}^2$. By adjoining to $\mathbb{R}^2$ a point at infinity and applying the Jordan curve theorem (see, for example, Wall (1972)) to the resulting sphere S^2, we see that $\mathbb{R}^2 \setminus g$ has exactly two connected components and g is their common boundary. We will call these two connected components the 'sides' of g.

If $a = g_{ab}(\alpha)$ and $b = g_{ab}(\beta)$, we define the segment $[a,b]$ to be the set of all points $c \in g_{ab}$ with $c = g_{ab}(\tau)$ for some τ such that $\alpha \leq \tau \leq \beta$ if $\alpha < \beta$, or such that $\beta \leq \tau \leq \alpha$ if $\beta < \alpha$. We also put $[a,a] = \{a\}$ for any $a \in \mathbb{R}^2$. It may be immediately verified that, with segments so defined, the axioms **(L1)–(L4)** are all satisfied. Thus we are free to use the results of Chapter III, Section 1, and g_{ab} is what is there denoted by <*a,b*>. This notation will now also be used here.

We now show that the axiom **(C)** is also satisfied. Without loss of generality, suppose $c \in (a,b_1)$ and $d \in (c,b_2)$. By Proposition III.6 we may assume that $b_2 \notin$ <*a,b*$_1$>. Then b_1 and b_2 are on opposite sides of <*a,d*>, since c and b_2 are on opposite sides of <*a,d*> but c and b_1 are on the same side. Thus there exists a point $b \in (b_1,b_2) \cap$ <*a,d*>.

Assume first that $b \in (a,d)$. Then c and d are on opposite sides of $\langle b_1,b_2\rangle$, since a and d are on opposite sides of $\langle b_1,b_2\rangle$ but a and c are on the same side. Since $d \in (c,b_2)$, this is a contradiction.

Assume next that $a \in (d,b)$. Then b and b_1 are on opposite sides of $\langle c,d\rangle$, since a and b_1 are on opposite sides of $\langle c,d\rangle$ but a and b are on the same side. Thus there exists a point $b' \in (b_1,b) \cap \langle c,d\rangle$. Since $b' \in (b_1,b_2)$ and $b_2 \in \langle c,d\rangle$, this is a contradiction.

We conclude that $d \in (a,b)$. Thus $\mathbb{R}^2$ is a convex geometry with the present definition of segments. Since the axiom **(U)** is evidently satisfied, it follows from Proposition V.26 that the axioms **(P)** and **(D)** are also satisfied.

As a specific example, consider the family $\mathcal{G}$ consisting of all subsets

$$g^{\alpha\beta} = \{x = (\xi_1,\xi_2) \in \mathbb{R}^2 \colon \xi_2 - \beta = (\xi_1 - \alpha)^4\}$$

and all subsets

$$h^{\gamma} = \{x = (\xi_1,\xi_2) \in \mathbb{R}^2 \colon \xi_1 = \gamma\},$$

where $\alpha,\beta,\gamma \in \mathbb{R}$. Each subset $g^{\alpha\beta}$ is a translate of $g = g^{00}$ and each subset h^{γ} is a translate of $h = h^0$. It follows that the family $\mathcal{G}$ satisfies the condition (i) above. We now show that $\mathcal{G}$ also satisfies the condition (ii).

Let $x = (\xi_1,\xi_2)$ and $y = (\eta_1,\eta_2)$ be arbitrary distinct points of $\mathbb{R}^2$. If $\xi_1 = \eta_1 = \delta$, then x and y are both contained in h^{δ}, but are not both contained in any other subset in $\mathcal{G}$. If $\xi_1 \neq \eta_1$, then x and y are not both contained in any subset h^{γ}. We wish to show that they are both contained in a unique subset $g^{\alpha\beta} \in \mathcal{G}$. We require

$$\xi_2 - \beta = (\xi_1 - \alpha)^4, \quad \eta_2 - \beta = (\eta_1 - \alpha)^4,$$

which implies

$$\eta_2 - \xi_2 = (\eta_1 - \alpha)^4 - (\xi_1 - \alpha)^4.$$

Since $\xi_1 \neq \eta_1$, this is a cubic equation for α with at least one real root. In fact it has a unique real root, since the polynomial in α on the right side has a nonvanishing derivative. Having determined α, we immediately obtain $\beta = \xi_2 - (\xi_1 - \alpha)^4 = \eta_2 - (\eta_1 - \alpha)^4$.

As another example, let $g = \{x = (\xi_1,\xi_2) \in \mathbb{R}^2 \colon \xi_2 > 0, \xi_1 = -\xi_2^{-1}\}$ be the branch of the hyperbola $\xi_1\xi_2 = -1$ in the upper half-plane. Given any positive numbers μ,ρ, there exist uniquely determined points $x = (\xi_1,\xi_2)$, $y = (\eta_1,\eta_2)$ on g, with $\xi_1 < \eta_1$, such that the straight line segment joining x and y has slope μ and length ρ. In fact, if we put $\sigma = \rho/2(1 + \mu^2)^{1/2}$, then

$$\xi_1 = -\sigma - (\sigma^2 + \mu^{-1})^{1/2}, \quad \eta_1 = \sigma - (\sigma^2 + \mu^{-1})^{1/2}.$$

Let

$$g^{\alpha\beta} = \{x = (\xi_1,\xi_2) \in \mathbb{R}^2 : (\xi_1 - \alpha, \xi_2 - \beta) \in g\},$$

where $\alpha,\beta \in \mathbb{R}$, be an arbitrary translate of g. Then it follows that, given any two distinct points $x = (\xi_1,\xi_2)$ and $y = (\eta_1,\eta_2)$ of $\mathbb{R}^2$ such that the straight line segment joining x and y has slope $\mu > 0$ and length $\rho > 0$, there is exactly one translate $g^{\alpha\beta}$ which contains both x and y; in fact we must take

$$\alpha = (\sigma^2 + \mu^{-1})^{1/2} + (\xi_1 + \eta_1)/2, \quad \beta = -\mu(\sigma^2 + \mu^{-1})^{1/2} + (\xi_2 + \eta_2)/2.$$

Hence the family $\mathcal{G}$ consisting of all subsets $g^{\alpha\beta}$ of $\mathbb{R}^2$, together with all straight lines in $\mathbb{R}^2$ which do not have positive slope,

$$h^{\gamma} = \{x = (\xi_1,\xi_2) \in \mathbb{R}^2 : \xi_1 = \gamma\}, \quad \ell^{\mu\delta} = \{x = (\xi_1,\xi_2) \in \mathbb{R}^2 : \xi_2 = \mu\xi_1 + \delta\},$$

where $\gamma,\delta,\mu \in \mathbb{R}$ and $\mu \leq 0$, satisfies the conditions (i) and (ii).

Further interesting examples of dense, unending linear geometries can be given, but they require an acquaintance with projective geometry. For this reason, and also for future reference, we now give a brief introduction to this subject.

2 PROJECTIVE GEOMETRY

A *projective space P* (in the terminology of some authors, an *irreducible* projective space) is a set of elements, called 'points', and a nonempty collection of nonempty proper subsets, called 'lines', such that

(i) any two distinct 'points' a,b are contained in exactly one 'line', denoted here by $\langle a,b \rangle$,
(ii) every 'line' contains at least three distinct 'points',
(iii) if $\langle a,b \rangle$ and $\langle c,d \rangle$ are distinct 'lines' with a common 'point' $e \neq a,b,c,d$, then $\langle a,d \rangle$ and $\langle b,c \rangle$ are also distinct 'lines' with a common 'point' $f \neq a,b,c,d$ (see Figure 2).

Two 'lines' will be said to intersect if they have a common 'point' and 'points' will be said to be collinear if they are all contained in a common 'line'. (Inverted commas are used for several concepts in projective space to distinguish them from the corresponding concepts in a vector space.) A *projective plane* is a projective space in which any two 'lines' intersect.

For example, with any vector space V over a division ring D there is associated a projective space $P(V)$ whose 'points' are the one-dimensional vector subspaces of V and whose 'lines' are the two-dimensional vector subspaces of V. If W is a vector subspace of V, then $P(W)$ is said to be a 'subspace' of $P(V)$, and if W is a maximal proper vector subspace of V, then $P(W)$ is said to be a 'hyperplane' of $P(V)$.

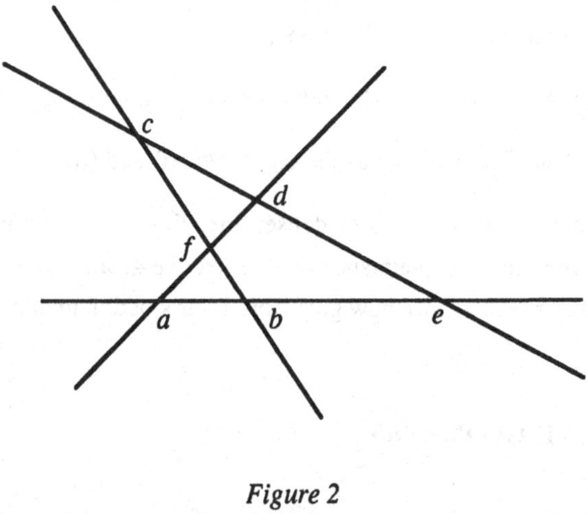

Figure 2

Let V be an arbitrary vector space over a division ring D and let $V^* = D \oplus V$ be the vector space of all couples (λ, v), where $\lambda \in D$ and $v \in V$, with the obvious definitions of addition and multiplication by a scalar:

$$(\lambda_1, v_1) + (\lambda_2, v_2) = (\lambda_1 + \lambda_2, v_1 + v_2), \quad \mu(\lambda, v) = (\mu\lambda, \mu v).$$

We can regard V as a subset of the projective subspace $P(V^*)$ by identifying the vector $v \in V$ with the one-dimensional subspace of V^* containing the couple $(1,v)$. The remaining 'points' of $P(V^*)$, i.e. the one-dimensional subspaces of V^* containing a couple $(0,v)$ for some $v \in V \setminus \{0\}$, form a 'hyperplane' of $P(V^*)$ – the 'hyperplane at infinity'. Thus $P(V^*)$ is the union of V and the 'hyperplane at infinity'. The projective space $P(V^*)$ is called the *projective completion* of V and will be denoted here by $\overline{V}$.

More concretely, if D is a division ring and n a positive integer, the 'points' of the n-dimensional projective space $P^n = P^n(D)$ may be represented by $(n + 1)$-tuples of elements of D, not all zero, with two $(n + 1)$-tuples $(\xi_0,\xi_1,\ldots,\xi_n)$ and $(\xi_0',\xi_1',\ldots,\xi_n')$ representing the same 'point' if $\xi_i' = \lambda\xi_i$ $(i = 0,1,\ldots,n)$ for some $\lambda \in D$. If x and y are distinct 'points', represented by the $(n + 1)$-tuples $(\xi_0,\xi_1,\ldots,\xi_n)$ and $(\eta_0,\eta_1,\ldots,\eta_n)$, then the 'points' of the 'line' $\langle x,y\rangle$ are represented by all $(n + 1)$-tuples $(\zeta_0,\zeta_1,\ldots,\zeta_n)$, where $\zeta_i = \lambda\xi_i + \mu\eta_i$ $(i = 0,1,\ldots,n)$ for some $\lambda,\mu \in D$, not both zero. The projective space P^n is the projective completion of the usual vector space of all n-tuples $(\xi_1,\ldots,\xi_n)$ of elements of D, the 'hyperplane at infinity' being the set of all 'points' of P^n which are represented by $(n + 1)$-tuples $(\xi_0,\xi_1,\ldots,\xi_n)$ with $\xi_0 = 0$.

Two distinct 'lines' of a projective space $P(V)$ are 'coplanar' if they are contained in a three-dimensional subspace of V. It follows at once from (i),(iii) that two distinct 'coplanar lines' of $P(V)$ intersect in a unique 'point'. An important attribute of the projective space $P(V)$ is Desargues' theorem:

If $\langle a_1,b_1\rangle,\langle a_2,b_2\rangle,\langle a_3,b_3\rangle$ *are three distinct 'lines' of* $P(V)$ *with a common 'point'* $p \neq a_i,b_i$ $(i = 1,2,3)$, *then the 'points' of intersection*

$$c_1 = \langle a_2,a_3\rangle \cap \langle b_2,b_3\rangle, \quad c_2 = \langle a_3,a_1\rangle \cap \langle b_3,b_1\rangle, \quad c_3 = \langle a_1,a_2\rangle \cap \langle b_1,b_2\rangle$$

are collinear.

We sketch the algebraic proof. Suppose the 'points' p,a_i,b_i of $P(V)$ are represented by the vectors π,α_i,β_i $(i = 1,2,3)$ of V. Since each is determined only up to a non-zero scalar multiple, we may assume that

$$\pi = \alpha_1 - \beta_1 = \alpha_2 - \beta_2 = \alpha_3 - \beta_3.$$

Then

$$\alpha_2 - \alpha_3 = \beta_2 - \beta_3, \quad \alpha_3 - \alpha_1 = \beta_3 - \beta_1, \quad \alpha_1 - \alpha_2 = \beta_1 - \beta_2.$$

Hence the 'points' c_1,c_2,c_3 are represented by the vectors $\alpha_2 - \alpha_3, \alpha_3 - \alpha_1$, $\alpha_1 - \alpha_2$. Since the three vectors have zero sum, they are contained in a two-dimensional subspace of V and the three 'points' are collinear.

Desargues' theorem implies its own converse (cf. the proof of Proposition 3 below):

If a_1,a_2,a_3 *and* b_1,b_2,b_3 *are non-collinear triples of 'points' in* $P(V)$ *such that*

$$c_1 = <a_2,a_3> \cap <b_2,b_3>, \quad c_2 = <a_3,a_1> \cap <b_3,b_1>, \quad c_3 = <a_1,a_2> \cap <b_1,b_2>$$

are distinct collinear 'points', then the 'lines' $<a_1,b_1>,<a_2,b_2>,<a_3,b_3>$ *have a common 'point'* p.

Suppose now that $P(V)$ is the projective space associated with a vector space V over an *ordered* division ring D. If a,b are distinct 'points' of $P(V)$, represented by the vectors α,β of V, then the 'line' $<a,b>$ is the disjoint union of the 'points' a,b and two *open segments*, represented by the vectors $\lambda\alpha + \mu\beta$, where $\lambda,\mu \in D$ and $\lambda\mu > 0$ for one segment, $\lambda\mu < 0$ for the other. The corresponding *closed segments* are obtained by adjoining the two *endpoints* a,b.

It is readily seen that if $<a_i,b_i>$ $(i = 1,2,3,4)$ are four distinct 'lines' with a common 'point' $p \neq a_i,b_i$ $(i = 1,2,3,4)$, and if $a_3,a_4 \in <a_1,a_2>$, $b_3,b_4 \in <b_1,b_2>$, then b_3,b_4 belong to the same open segment with endpoints b_1,b_2 of the 'line' $<b_1,b_2>$ if a_3,a_4 belong to the same open segment with endpoints a_1,a_2 of the 'line' $<a_1,a_2>$.

EXAMPLE 6 The projective plane X over the field $\mathbb{Q}$ of rational numbers may be given the structure of a dense, unending linear geometry in the following way. Regard X as a subset of the real projective plane X' and choose a 'line' λ' in X' which contains no point of X. (For example, if $\alpha_1,\alpha_2 \in \mathbb{R}$ are such that $1,\alpha_1,\alpha_2$ are linearly independent over $\mathbb{Q}$, then $\alpha_1\alpha_2 \neq 0$ and we can take λ' to be the 'line' determined by the 'points' $(1,\alpha_1^{-1},0)$ and $(1,0,\alpha_2^{-1})$.) Any two points $x,y \in X$ with $x \neq y$ determine a unique 'line' $\ell' \subseteq X'$, and ℓ' intersects λ' in a unique point $z' \notin X$. The points of ℓ' other than x and y form two disjoint open segments. We define the segment $[x,y]$ of X to be the intersection with X of the open segment of ℓ' which does *not* contain z', together with x and y.

Once again, the only axiom which needs to be verified is (C). One of the configurations which can arise in the course of this verification is illustrated in Figure 3.

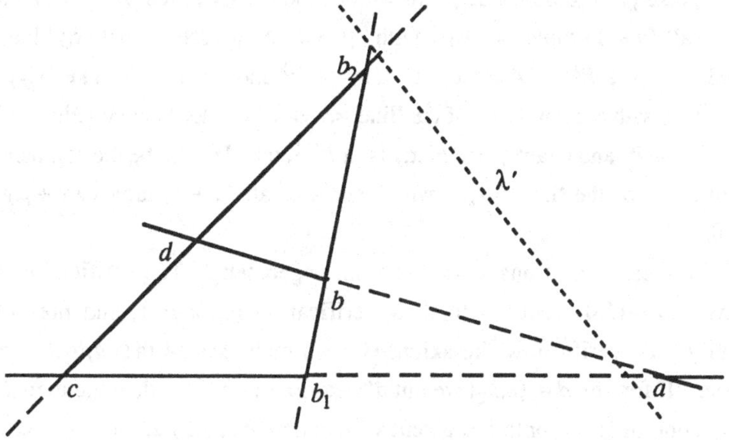

Figure 3: Veblen's projective plane

This example of a dense, unending linear geometry, which was given by Veblen (1904), may be readily generalized. Let D be an ordered division ring and suppose D is a subring of an ordered division ring D' which has dimension $> n$, considered as a vector space over D. It will be shown in Chapter VIII that such a division ring D' exists, for any positive integer n, if D is not isomorphic to $\mathbb{R}$. Then we can take X to be the n-dimensional projective space over D, and λ' to be a 'hyperplane' in the enclosing projective space X' over D' which contains no point of X. (For example, if $\alpha_1,\ldots,\alpha_n \in D'$ are such that $1,\alpha_1,\ldots,\alpha_n$ are linearly independent over D, we can take λ' to be the 'hyperplane' determined by the 'points' $(1,\alpha_1^{-1},0,\ldots,0),(1,0,\alpha_2^{-1},\ldots,0),\ldots,(1,0,0,\ldots,\alpha_n^{-1})$.)

EXAMPLE 7 For any ordered division ring D and any positive integer n, the n-dimensional projective space $P^n = P^n(D)$ can be given the structure of a dense, but not unending, linear geometry in the following way. We regard P^n as the set of $(n+1)$-tuples of elements of D, not all zero, with proportional $(n+1)$-tuples identified, and P^j $(0 \leq j < n)$ as the set of all $(n+1)$-tuples $(\xi_0,\xi_1,\ldots,\xi_n)$ with $\xi_0 = \xi_1 = \ldots = \xi_{n-j-1} = 0$. In particular, P^0 contains the unique 'point' $p_0 = (0,\ldots,0,\xi_n)$.

For any $x,y \in P^n$ with $x \neq y$ put $[x,x] = \{x\}$ and define $[x,y]$ inductively in the following way. If $x,y \in P^1$ and $x,y \neq p_0$, take $[x,y]$ to be the segment with endpoints x,y of the 'line' $\langle x,y\rangle$ which does not contain p_0; if $x,y \in P^1$ and $x = p_0$ or $y = p_0$ take $[x,y]$ to be the segment with endpoints x,y of the 'line' $\langle x,y\rangle$ which contains all $(n + 1)$-tuples $\lambda x + \mu y$ with $\lambda\mu > 0$. Now assume that $[x,y]$ has been defined for $x,y \in P^{k-1}$, where $k > 1$. If $x,y \in P^k$ and $x,y \notin P^{k-1}$, take $[x,y]$ to be the segment with endpoints x,y of the 'line' $\langle x,y\rangle$ which does not contain a point of P^{k-1}; if $x,y \in P^k$ and exactly one of x,y is in P^{k-1}, take $[x,y]$ to be the segment with endpoints x,y of the 'line' $\langle x,y\rangle$ which contains all $(n + 1)$-tuples $\lambda x + \mu y$ with $\lambda\mu > 0$.

With these definitions $X = P^n$ is a linear geometry. The verification of the axioms **(L2)–(L4)** reduces to their verification for $n = 1$, and presents no difficulty. To verify now the axiom **(C)** we may assume that a,b_1,b_2 are not collinear. If for any $d \in [c,b_2]$ we put $d' = \langle a,d\rangle \cap \langle b_1,b_2\rangle$, then the map $d \to d'$ is a bijection of $[c,b_2]$ onto a segment with endpoints b_1,b_2 of the 'line' $\langle b_1,b_2\rangle$. By considering separately the various possibilities, it may be seen that in all cases this segment is $[b_1,b_2]$. Since X is obviously dense, the axiom **(P)** follows from Proposition V.1. However, the dense linear geometry X is not unending, since if $b \neq p_0$ there is no $c \neq p_0$ such that $p_0 \in [b,c]$.

3 THE DESARGUES PROPERTY

A linear geometry will be said to have the *Desargues property* if the following condition (illustrated in Figure 4) is satisfied:

(Δ) *Let* $\langle a_1,b_1\rangle, \langle a_2,b_2\rangle, \langle a_3,b_3\rangle$ *be three distinct lines with a common point* $p \neq a_i,b_i$ $(i = 1,2,3)$. *If the lines* $\langle a_j,a_k\rangle$ *and* $\langle b_j,b_k\rangle$ *intersect for* $(j,k) = (1,2),(1,3),(2,3)$, *then the points of intersection* c_3,c_2,c_1 *(which are necessarily uniquely determined and distinct) are collinear.*

The Desargues property is vacuous in a one-dimensional linear geometry. It will now be shown that it need not hold in a dense unending two-dimensional linear geometry. (The qualification 'dense' is redundant, by Proposition V.26.) In Example 4 take $\mu = 4/9$ and $a_1 = (0,4)$, $a_2 = (6,2)$, $a_3 = (4,2)$, $b_1 = (8,4)$, $b_2 = (10,-2)$, $b_3 = (4,-1)$. Then the lines $\langle a_1,b_1\rangle, \langle a_2,b_2\rangle, \langle a_3,b_3\rangle$ have the common point $p = (4,4)$. The lines $\langle a_1,a_2\rangle$ and $\langle b_1,b_2\rangle$ intersect in the point $c_3 = (9,1)$, the lines $\langle a_2,a_3\rangle$ and $\langle b_2,b_3\rangle$ intersect in the point $c_1 = (-14,2)$, and the lines $<a_1,a_3>$ and $<b_1,b_3>$ intersect in the point $c_2 = (160/29, 36/29)$. However, c_2 does not lie on the line $<c_1,c_3>$. (Only one of the lines mentioned, namely $<b_1,b_3>$, is a 'refracted' line. Since the ordinary affine plane has the Desargues property, the refracted plane cannot.)

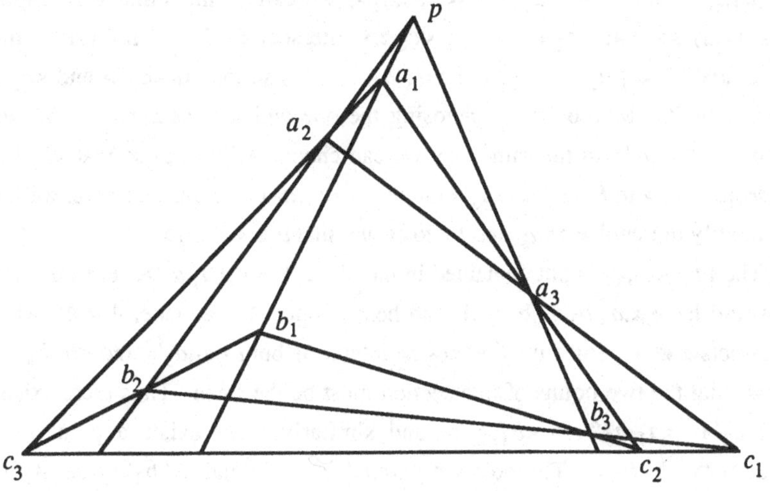

Figure 4: The Desargues property

In view of this example it is rather remarkable that the Desargues property must hold in a dense linear geometry which is not two-dimensional:

PROPOSITION 1 *If X is a dense linear geometry with* $\dim X > 2$, *then X has the Desargues property.*

Proof Let p, a_i, b_i, c_i $(i = 1,2,3)$ be points satisfying the hypotheses of (Δ). We suppose first that the affine set $A = <p,a_1,a_2,a_3>$ is three-dimensional. Then $<a_1,a_2,a_3>$ and $<b_1,b_2,b_3>$ are distinct planes in A with the common points c_1,c_2,c_3. Since the intersection of the two planes is affine, it must be a line. Thus from now on we may suppose that A is a plane.

Choose $\bar{x} \in X \setminus A$ and then choose $x \in P^i$, where P is the polytope $[p,a_1,a_2,a_3,b_1,b_2,b_3,c_1,c_2,c_3,\bar{x}]$. Since $x \notin <p,b_3> = <a_3,b_3>$, we can choose $y \in (p,x)$ so that $<a_3,x>$ and $<b_3,y'>$ intersect within P^i for any $y' \in (x,y]$. Indeed if $a_3 \in (p,b_3)$ or $p \in (a_3,b_3)$ we can choose any $y \in (p,x)$, whereas if $b_3 \in (p,a_3)$ we can choose $x' \in P^i$ so that $x \in (a_3,x')$ and then choose $y \in (p,x) \cap (b_3,x')$.

Similarly we can choose $y \in P^i$ with $x \in (p,y)$ so that $\langle a_3,x\rangle$ and $\langle b_3,y'\rangle$ intersect within P^i for any $y' \in (x,y]$.

Choose such a $y \in (p,x)$ if $b_3 \in (p,a_3)$, and such a y with $x \in (p,y)$ if $a_3 \in (p,b_3)$ or $p \in (a_3,b_3)$. If $z \in \langle a_3,x\rangle \cap \langle b_3,y\rangle$, then $z \in P^i$ and in every case $x \in (a_3,z)$. Since $z \notin \langle a_3,a_1\rangle = \langle c_2,a_1\rangle$, we can in the same way choose $a_3' \in (x,z)$ so that $\langle a_1,a_3'\rangle$ and $\langle c_2,z\rangle$ intersect in P^i. Similarly, since $z \notin \langle a_3,a_2\rangle = \langle c_1,a_2\rangle$, we can choose $a_3'' \in (x,z)$ so that $\langle a_2,a_3''\rangle$ and $\langle c_1,z\rangle$ intersect in P^i. Moreover, by choosing the one which is nearer to z, we may suppose $a_3'' = a_3'$. In the same way we can choose $b_3' \in (y,z)$ so that $\langle b_1,b_3'\rangle$ intersects $\langle c_2,z\rangle$ in P^i and $\langle b_2,b_3'\rangle$ intersects $\langle c_1,z\rangle$ in P^i. Furthermore, we may now modify our choice of a_3' and b_3' to ensure that $p \in \langle a_3',b_3'\rangle$.

The line $\langle c_2,z\rangle$ is not contained in the plane $A' = \langle p,a_1,a_3'\rangle$. For otherwise we would have $x,a_3,b_3,a_1,b_1 \in A'$ and hence, since A is a plane, $A = A'$, which contradicts $x \notin A$. Since the line $\langle c_2,z\rangle$ intersects both $\langle a_1,a_3'\rangle$ and $\langle b_1,b_3'\rangle$, it follows that the two points of intersection must be the same. Thus there exists a point $c_2' \in \langle a_1,a_3'\rangle \cap \langle b_1,b_3'\rangle$, and similarly there exists a point $c_1' \in \langle a_2,a_3'\rangle \cap \langle b_2,b_3'\rangle$. The points p,a_i,b_i,c_i' $(i = 1,2)$ and a_3',b_3',c_3 satisfy the hypotheses of (Δ) and the points p,a_1,a_2,a_3' are not coplanar, since $a_2 \notin A'$. Hence, by the first part of the proof, c_1',c_2',c_3 are collinear. Thus $c_3 \in \langle c_1,c_2,z\rangle$. Since $z \notin A$ and $\langle c_1,c_2,c_3\rangle \subseteq A$, it now follows from Proposition III.8 that $c_3 \in \langle c_1,c_2\rangle$. □

Although the Desargues property does not hold in all two-dimensional dense linear geometries, Proposition 1 shows that it must hold in such a geometry if it can be embedded in a dense linear geometry of higher dimension. Consequently *we will assume throughout the remainder of this section that X is a dense linear geometry of dimension* ≥ 2 *with the Desargues property.*

We show first that a modified form of the Desargues property also holds:

PROPOSITION 2 *Let* $\langle a_1,b_1\rangle, \langle a_2,b_2\rangle, \langle a_3,b_3\rangle$ *be three distinct lines with a common point* $p \neq a_i,b_i$ $(i = 1,2,3)$. *If*

$$c_1 = \langle a_2,a_3\rangle \cap \langle b_2,b_3\rangle,\ c_2 = \langle a_3,a_1\rangle \cap \langle b_3,b_1\rangle,\ c_3 = \langle a_1,a_2\rangle \cap \langle c_1,c_2\rangle,$$

then $c_3 \in \langle b_1,b_2\rangle$.

Proof The hypotheses imply that $c_1 \neq a_2,a_3,b_2,b_3$ and $c_2 \neq a_1,a_3,b_1,b_3$. Moreover $c_1 \neq c_2$. Hence the lines $\langle a_1,c_2\rangle$, $\langle a_2,c_1\rangle$, $\langle p,b_3\rangle$ are distinct and have the common point a_3. It now follows from (Δ) that $c_3 \in \langle b_1,b_2\rangle$. □

We establish next a converse of the Desargues property:

PROPOSITION 3 *Let a_1,a_2,a_3 and b_1,b_2,b_3 be two non-collinear triples of points such that*

$$c_1 = \langle a_2,a_3\rangle \cap \langle b_2,b_3\rangle,\ c_2 = \langle a_3,a_1\rangle \cap \langle b_3,b_1\rangle,\ c_3 = \langle a_1,a_2\rangle \cap \langle b_1,b_2\rangle$$

exist and are distinct collinear points. If $p \in \langle a_1,b_1\rangle \cap \langle a_2,b_2\rangle$, then also $p \in \langle a_3,b_3\rangle$.

Proof Since the points c_1,c_2,c_3 are distinct, we have $a_i \neq b_i$ $(i = 1,2,3)$. Since a_1,a_2,a_3 and b_1,b_2,b_3 are not collinear, we also have $c_i \neq a_i,b_i$ $(i = 1,2,3)$. We may further assume that $c_1,c_2 \neq a_1,a_2,b_1,b_2$. For suppose $c_2 = a_1$. Then $a_2 \in \langle c_2,c_3\rangle = \langle c_1,c_2\rangle$. Since a_1,a_2,a_3 do not all lie on the line $\langle c_1,c_2\rangle$, it follows that $c_1 = a_2$. Thus $a_1 \in \langle b_1,b_3\rangle$, $a_2 \in \langle b_2,b_3\rangle$, and hence $p = b_3$.

With this assumption the lines $\langle a_1,a_2\rangle$, $\langle b_1,b_2\rangle$, $\langle c_2,c_1\rangle$ are distinct, and by hypothesis they have the common point c_3. We may assume in addition that $c_3 \neq a_1,a_2,b_1,b_2$. For suppose $c_3 = a_1$. Then $a_3 \in \langle c_2,c_3\rangle = \langle c_1,c_3\rangle$ and hence $c_1 = a_3$, since a_1,a_2,a_3 are not collinear. Moreover $\langle a_1,b_1\rangle = \langle b_1,b_2\rangle$ and hence $p = b_2$. Since $c_1 \in \langle b_2,b_3\rangle$ and $c_1 = a_3 \neq b_3$, we obtain $p \in \langle a_3,b_3\rangle$.

With this additional assumption we have

$$a_3 = \langle a_1,c_2\rangle \cap \langle a_2,c_1\rangle,\ b_3 = \langle b_1,c_2\rangle \cap \langle b_2,c_1\rangle,\ p = \langle a_1,b_1\rangle \cap \langle a_2,b_2\rangle$$

and hence, by (Δ), $p \in \langle a_3,b_3\rangle$. □

The Desargues property and its converse may be loosely summarized by saying that 'two triangles are perspective from a point if and only if they are perspective from a line'.

In the next chapter we will make much use of the following result. Since under our hypotheses two coplanar lines need not always intersect, the proof of the result will be a good deal more complicated than that of its counterpart in projective geometry.

PROPOSITION 4 *Let a_1,a_2,a_3,a_4 and a_1',a_2',a_3',a_4' be two quadruples of points, such that no three points of the same quadruple are collinear and none of the points lies on a given line ℓ. If $<a_j,a_k>$ and $<a_j',a_k'>$ intersect ℓ in the same point a_{jk}, for $(j,k) = (1,2),(1,3),(1,4),(2,3),(3,4)$, and if $<a_2,a_4>$ intersects ℓ in a_{24}, then $<a_2',a_4'>$ also intersects ℓ in a_{24}.*

Proof Since a_1,a_2,a_3 are not collinear, we must have $a_{12} \neq a_{13}$. Similarly $a_{12},a_{13} \neq a_{14},a_{23}$ and $a_{34} \neq a_{13},a_{14},a_{23}$. The set-up is illustrated in Figure 5.

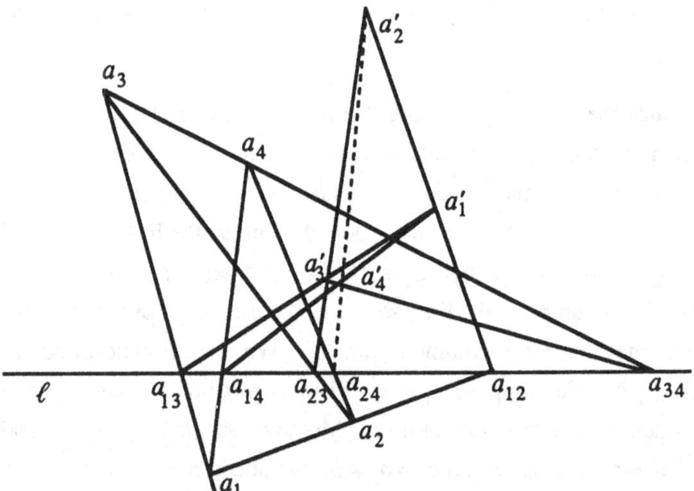

Figure 5

We consider first the case in which the two quadruples have a corresponding point in common. Since $a_j = a_j'$, $a_k = a_k'$ for some $j,k \in \{1,2,3,4\}$ with $j \neq k$ implies $a_i = a_i'$ for all $i \in \{1,2,3,4\}$, we need only consider the case in which $a_i = a_i'$ for exactly one $i \in \{1,2,3,4\}$. If $a_1 = a_1'$, then the lines $<a_2,a_2'>$, $<a_4,a_4'>$, $<a_3,a_3'>$ have the common point a_1 and hence, by Proposition 2, $a_{24} \in <a_2',a_4'>$. If $a_3 = a_3'$, then the lines $<a_4,a_4'>,<a_2,a_2'>,<a_1,a_1'>$ have the common point a_3 and hence, in the same way, $a_{24} \in <a_2',a_4'>$. If $a_2 = a_2'$ we obtain $a_{24} \in <a_2',a_4'>$ by applying Proposition 3 to the triples a_3,a_1,a_4 and

a_3', a_1', a_4'. If $a_4 = a_4'$ we arrive at the same conclusion by applying Proposition 3 to the triples a_1, a_3, a_2 and a_1', a_3', a_2'. Thus we may now assume $a_i \neq a_i'$ ($i = 1,2,3,4$).

Assume first that $\langle a_j, a_k \rangle \neq \langle a_j', a_k' \rangle$ for $(j,k) = (1,2),(1,3),(1,4),(2,3),(3,4)$. Suppose also that there exists a point $q \in \langle a_1, a_1' \rangle \cap \langle a_3, a_3' \rangle$. Since a_1, a_3, a_2 and a_1', a_3', a_2' are non-collinear triples, and since a_{23}, a_{12}, a_{13} are distinct collinear points, it follows from Proposition 3 that $q \in \langle a_2, a_2' \rangle$. In the same way, since a_1, a_3, a_4 and a_1', a_3', a_4' are non-collinear triples, and since a_{34}, a_{14}, a_{13} are distinct collinear points, $q \in \langle a_4, a_4' \rangle$. Thus $q = \langle a_2, a_2' \rangle \cap \langle a_4, a_4' \rangle$. Since a_2, a_{23}, a_2' and a_4, a_{34}, a_4' are non-collinear triples, and since a_3', q, a_3 are distinct collinear points, it follows from Proposition 3 again that $a_{24} \in \langle a_2', a_4' \rangle$, as we wished to show. Thus we may now suppose that $\langle a_1, a_1' \rangle \cap \langle a_3, a_3' \rangle = \emptyset$. Then $a_{13} \in (a_1', a_3')$ if $a_{13} \in (a_1, a_3)$, $a_1' \in (a_3', a_{13})$ if $a_1 \in (a_3, a_{13})$, and $a_3' \in (a_1', a_{13})$ if $a_3 \in (a_1, a_{13})$.

We now drop the assumption made at the beginning of the previous paragraph. Our strategy will be to construct from the quadruple a_1, a_2, a_3, a_4 another quadruple b_1, b_2, b_3, b_4, no three points of which are collinear, with $\langle b_j, b_k \rangle \neq \langle a_j', a_k' \rangle$ and $a_{jk} \in \langle b_j, b_k \rangle$ for $(j,k) = (1,2),(1,3),(1,4),(2,3),(3,4)$, with $a_{24} \in \langle b_2, b_4 \rangle$, and such that a_{13}, b_1, b_3 are *not* in the same order as a_{13}, a_1', a_3'.

By interchanging a_1 and a_3, if necessary, we may assume that $a_{34} \notin (a_{13}, a_{14})$. Choose any $b_1 \in (a_1, a_{12})$. Put $d = a_{13}$ if $a_{12} \in (a_{13}, a_{23})$ or $a_{13} \in (a_{12}, a_{23})$, and put $d = (a_1, a_{23}) \cap (b_1, a_{13})$ if $a_{23} \in (a_{12}, a_{13})$. Then for any $b_3 \in (b_1, d)$ there exists $b_2 \in (a_1, a_{12}) \setminus b_1$ such that $a_{23} \in \langle b_2, b_3 \rangle$. If $a_{14} \in (a_{13}, a_{34})$ take $b_4 = (b_1, a_{14}) \cap (b_3, a_{34})$, and if $a_{13} \in (a_{14}, a_{34})$ take $b_4 \in (b_1, a_{14})$ so that $b_3 \in (b_4, a_{34})$.

By construction no three points of the quadruple b_1, b_2, b_3, b_4 are collinear and $a_{jk} \in \langle b_j, b_k \rangle$ for $(j,k) = (1,2),(1,3),(1,4),(2,3),(3,4)$. Moreover $b_1 \in (a_1, a_{12})$, $b_3 \in (b_1, a_{13})$ and $b_2 \in (a_1, a_{12}) \setminus b_1$. We are going to deduce from these properties that also $a_{24} \in \langle b_2, b_4 \rangle$.

Put $d' = a_{13}$ if $a_{13} \in (a_1, a_3)$ or $a_1 \in (a_3, a_{13})$, and put $d' = (a_3, a_{12}) \cap (b_1, a_{13})$ if $a_3 \in (a_1, a_{13})$. Then every line through a_3 which intersects (b_1, d') also intersects $(a_1, a_{12}) \setminus b_1$. Put $d'' = a_{13}$ if $a_{12} \in (a_{13}, a_{23})$ or $a_{13} \in (a_{12}, a_{23})$, and put $d'' = (a_1, a_{23}) \cap (b_1, a_{13})$ if $a_{23} \in (a_{12}, a_{13})$. Then every line through a_{23} which intersects (b_1, d'') also intersects $(a_1, a_{12}) \setminus b_1$. We may now choose $c_3 \in (b_1, d') \cap (b_1, d'') \cap (b_1, b_3)$ so that $c_3 \notin \langle a_2, a_3 \rangle$, $c_3 \notin \langle a_3, a_4 \rangle$, the line

$<c_3,a_3>$ intersects $(a_1,a_{12})\setminus b_1$ in a point p such that $b_2 \notin [b_1,p]$, and the line $<c_3,a_{23}>$ intersects $(a_1,a_{12})\setminus b_1$ in a point c_2 such that $b_2 \notin [b_1,c_2]$. Evidently b_1,c_2,c_3 are not collinear. If $a_{14} \in (a_{13},a_{34})$ take $c_4 = (b_1,a_{14}) \cap (c_3,a_{34})$, and if $a_{13} \in (a_{14},a_{34})$ take $c_4 \in (b_1,a_{14})$ so that $c_3 \in (c_4,a_{34})$. In either case b_1,c_3,c_4 are not collinear and $c_4 = (b_1,a_{14}) \cap <c_3,a_{34}>$. Indeed if we put $c_1 = b_1$, then no three points of the quadruple c_1,c_2,c_3,c_4 are collinear and $a_{jk} \in <c_j,c_k>$ for $(j,k) = (1,2),(1,3),(1,4),(2,3),(3,4)$. Furthermore $c_i \neq b_i$, not only for $i = 2,3$ but also for $i = 4$, since

$$<b_1,a_{13}> \cap <c_4,a_{34}> = c_3 \neq b_3 = <b_1,a_{13}> \cap <b_4,a_{34}>.$$

Since a_{23},a_{13},a_{12} are distinct collinear points and $p \in <c_1,a_1> \cap <c_3,a_3>$, by applying Proposition 3 to the triples c_1,c_3,c_2 and a_1,a_3,a_2 we obtain $p \in <c_2,a_2>$. Similarly by applying Proposition 3 to the triples c_1,c_3,c_4 and a_1,a_3,a_4 we obtain $p \in <c_4,a_4>$. Thus

$$p \in <c_i,a_i> \quad (i = 1,2,3,4).$$

Moreover $p \neq c_i,a_i$ $(i = 1,2,3,4)$. For $b_1 \notin <a_1,a_3>$ implies $p \neq b_1,a_1,c_3,a_3$. Similarly $c_3 \notin <a_2,a_3>$ implies $p \neq a_2,c_2$ and $c_3 \notin <a_3,a_4>$ implies $p \neq a_4,c_4$.

The triple a_2,a_{23},c_2 is not collinear, since $c_3 \notin <a_2,a_3>$, and the triple a_4,a_{34},c_4 is not collinear, since $c_3 \notin <a_3,a_4>$. Since c_3,p,a_3 are distinct collinear points and $a_{24} \in <a_2,a_4> \cap \ell$ it follows from Proposition 3 again that $a_{24} \in <c_2,c_4>$. Since $c_2 \in <b_1,b_2>$ and $c_2 \neq b_2$, we have $c_2 \notin <b_2,b_3>$. Thus the points c_2,a_{23},b_2 are not collinear, and similarly the points c_4,a_{34},b_4 are not collinear. Applying Proposition 3 to these two triples we now obtain $a_{24} \in <b_2,b_4>$, since b_3,b_1,c_3 are distinct collinear points and $a_{24} \in <c_2,c_4> \cap \ell$.

On account of the arbitrary nature of b_1 and b_3, we may suppose that $<b_j,b_k> \neq <a_j',a_k'>$ for $(j,k) = (1,3),(1,4),(2,3),(3,4)$. However, if $<a_1,a_2> = <a_1',a_2'>$, we will have $<b_1,b_2> = <a_1',a_2'>$. We now apply to the quadruple b_1,b_2,b_3,b_4 a similar procedure to that which we applied to the quadruple a_1,a_2,a_3,a_4.

Choose any $b_1' \in (b_3,a_{13})$. If $a_{12} \in (a_{13},a_{23})$ or $a_{13} \in (a_{12},a_{23})$ put $e = a_{13}$, and if $a_{23} \in (a_{12},a_{13})$ put $e = (b_1',a_{13}) \cap (g,a_{23})$, with $g = (a_1,a_{13}) \cap <b_1',a_{12}>$. Then for any $b_3' \in (b_1',e)$, there exists $b_2' \in (g,a_{12})\setminus b_1'$ such that $a_{23} \in <b_2',b_3'>$. Now define b_4' in the analogous way to b_4. By construction no three points of the quadruple b_1',b_2',b_3',b_4' are collinear and $a_{jk} \in <b_j',b_k'>$ for $(j,k) = (1,2),(1,3),(1,4),(2,3),(3,4)$. In the same way as before we may show that

$a_{24} \in \langle b_2',b_4'\rangle$. On account of the arbitrary nature of b_1',b_3' we may suppose that $\langle b_j',b_k'\rangle \neq \langle a_j',a_k'\rangle$ for $(j,k) = (1,2),(1,4),(2,3),(3,4)$, but in addition $\langle b_1',b_3'\rangle = \langle b_1,b_3\rangle \neq \langle a_1',a_3'\rangle$. The quadruple b_1',b_2',b_3',b_4' has $b_3' \in (b_1',a_{13})$. If $a_3' \notin (a_1',a_{13})$, we are finished. To complete the proof we now construct another quadruple b_1'',b_2'',b_3'',b_4'' with the same properties but with $b_1'' \in (b_3'',a_{13})$.

We have $b_3' \in (b_1,a_{13})$ and $b_1 \in (a_1,a_{12})$. Choose any $b_1'' \in (b_3',a_{13})$. We are going to choose $y \in (b_1'',b_3')$ so that, for each $x \in (b_1'',y)$, the line $\langle x,a_{34}\rangle$ intersects the line $\langle b_1'',a_{14}\rangle$. If $a_{13} \in (a_{14},a_{34})$, choose any $y \in (b_1'',b_3')$. If $a_{14} \in (a_{13},a_{34})$ we choose y in the following way. If $a_{13} \in (a_{12},a_{34})$, then $b_1'' \in (u,a_{14})$ for some $u \in (b_1,a_{12})$ and we take $y' = (b_1,a_{13}) \cap (u,a_{34})$. If $a_{13} \notin (a_{12},a_{34})$, then $b_1'' \in (v,a_{14})$ for some $v \in (a_1,a_{13})$ and we take $y' \in (v,a_{34})$ so that $b_1'' \in (y',a_{13})$. Then in either of the last two cases we choose any $y \in (b_1'',b_3') \cap (b_1'',y')$.

If $a_{12} \in (a_{13},a_{23})$, take $z \in (a_1,a_{13})$ so that $b_1'' \in (z,a_{12})$ and then choose $z'' \in (b_1'',z)$ so that $y'' = (z'',a_{23}) \cap (b_1'',b_1)$ satisfies $y'' \in (b_1'',y)$. Choose any $b_2'' \in (b_1'',z'')$ and then take $b_3'' = (b_1'',y'') \cap (b_2'',a_{23})$.

If $a_{13} \in (a_{12},a_{23})$, take $z \in (y,a_{23})$ so that $b_1'' \in (z,a_{12})$, and if $a_{23} \in (a_{12},a_{13})$, take $z = (b_1'',a_{12}) \cap (y,a_{23})$. In either case choose any $b_2'' \in (b_1'',z)$ and then take $b_3'' \in (b_1'',y)$ so that $b_2'' \in (b_3'',a_{23})$.

In every case we now take $b_4'' = \langle b_3'',a_{34}\rangle \cap \langle b_1'',a_{14}\rangle$. By construction no three points of the quadruple b_1'',b_2'',b_3'',b_4'' are collinear, $b_1'' \in (b_3'',a_{13})$ and $a_{jk} \in \langle b_j'',b_k''\rangle$ for $(j,k) = (1,2),(1,3),(1,4),(2,3),(3,4)$. In the same way as before it follows that $a_{24} \in \langle b_2'',b_4''\rangle$ and we may choose b_1'',b_2'' so that $\langle b_j'',b_k''\rangle \neq \langle a_j',a_k'\rangle$ not only for $(j,k) = (1,3)$, but also for $(j,k) = (1,2),(1,4),(2,3),(3,4)$. □

4 FACTOR GEOMETRIES

In this section we study the way in which the Desargues property, and also the properties of density and unendingness considered in Chapter V, are inherited by the factor geometries of a linear geometry which were defined in Chapter IV.

PROPOSITION 5 *Let X be a linear geometry, H_+ an open half-space associated with a hyperplane H of X, and L a nonempty affine subset of H. If the linear geometry X is dense (resp. unending), then the linear geometry $H_+ : L$ is also dense (resp. unending).*

Proof Suppose $\rho_1, \rho_2 \in H_+ : L$ and $\rho_1 \neq \rho_2$. Choose $a_i \in \rho_i \backslash L$ $(i = 1,2)$. Then $a_1 \neq a_2$. If X is dense, there exists $a \in (a_1, a_2)$. Then $a \in H_+$ and $\rho := \langle a \cup L \rangle \cap (H_+ \cup H)$ satisfies $\rho \in (\rho_1, \rho_2)$.

If X is unending, there exists $a_3' \in X$ such that $a_1 \in (a_2, a_3')$. If $a_3' \in H_+$, put $a_3 = a_3'$. If $a_3' \notin H_+$ then, by Propositions III.12 and V.26, there exists a point $a_3 \in (a_1, a_3') \cap H_+$. Then $a_1 \in (a_2, a_3)$ and $\rho_3 := \langle a_3 \cup L \rangle \cap (H_+ \cup H)$ satisfies $\rho_1 \in (\rho_2, \rho_3)$. □

We next investigate the relationship between the dimensions of a linear geometry and its factor geometries. We first observe that if $\rho_1, \ldots, \rho_m \in H_+ : L$, and if $a_i \in \rho_i \backslash L$ $(i = 1, \ldots, m)$, then $[\rho_1, \ldots, \rho_m]$ is the set of all $\rho \in H_+ : L$ such that $\rho \cap [a_1, \ldots, a_m] \neq \emptyset$. This follows by induction from Lemma IV.24 (without any extra hypotheses).

Now let X be a dense linear geometry, H_+ an open half-space of X associated with a hyperplane H, and L a nonempty affine subset of H. Also let $a^0 \in H_+$, let $B = \{b^i : i \in I\}$ be an affine basis for L, and let $C = \{c^j : j \in J\}$ be an affine independent set (with $0 \notin J$ and possibly $J = \emptyset$) such that $B \cup C$ is an affine basis for H. For each $j \in J$, choose $a^j \in (a^0, c^j)$ and put $J' = J \cup \{0\}$, $A = \{a^j : j \in J'\}$. Then $A \cup B$ is an affine basis for X.

For each $j \in J'$, put $\rho^j = \langle a^j \cup L \rangle \cap (H_+ \cup H)$. We are going to show that *$R = \{\rho^j : j \in J'\}$ is an affine basis for $H_+ : L$.*

We show first that R affinely generates $H_+ : L$. Let $\rho \in H_+ : L$ and choose $a \in \rho \backslash L$. Since $A \cup B$ is an affine basis for X, there exist finite sets $\{a_1, \ldots, a_m\} \subseteq A$ and $\{b_1, \ldots, b_n\} \subseteq B$ such that $a \in \langle a_1, \ldots, a_m, b_1, \ldots, b_n \rangle$. Hence, by Proposition III.7, $a \in \langle c', c'' \rangle$ for some $c', c'' \in [a_1, \ldots, a_m, b_1, \ldots, b_n]$. Moreover we may choose the notation so that $c' \notin L$. Then $c' \in [a', b')$ for some $a' \in [a_1, \ldots, a_m]$ and $b' \in [b_1, \ldots, b_n]$. Hence $\rho' = \langle c' \cup L \rangle \cap (H_+ \cup H)$ satisfies $\rho' \in [\rho_1, \ldots, \rho_m]$, where $\rho_i = \langle a_i \cup L \rangle \cap (H_+ \cup H)$ $(i = 1, \ldots, m)$. If $c'' \in L$, then $\rho = \rho'$; if $c'' \notin L$ then, in the same way, $\rho'' = \langle c'' \cup L \rangle \cap (H_+ \cup H)$ satisfies $\rho'' \in [\rho_1, \ldots, \rho_m]$. Hence $\rho \in \langle \rho', \rho'' \rangle \subseteq \langle \rho_1, \ldots, \rho_m \rangle$.

It remains to show that R is affine independent. It is enough to show that if $\{\rho_1,...,\rho_m\}$ is an affine independent subset of R and if $\rho_{m+1} \in R$ is such that $\rho_{m+1} \in <\rho_1,...,\rho_m>$, then $\rho_{m+1} = \rho_i$ for some $i \le m$. We have $\rho_{m+1} \in <\rho',\rho''>$, where $\rho',\rho'' \in [\rho_1,...,\rho_m]$. Thus if $\rho_i = <a_i \cup L> \cap (H_+ \cup H)$, where $a_i \in A$ $(i = 1,...,m+1)$, then there exist $a' \in \rho'$, $a'' \in \rho''$ such that $a',a'' \in [a_1,...,a_m]$. If $\rho_{m+1} \in [\rho',\rho'']$, there exists $\tilde{a}_{m+1} \in \rho_{m+1}$ such that $\tilde{a}_{m+1} \in [a',a''] \subseteq [a_1,...,a_m]$ and hence $a_{m+1} \in <\tilde{a}_{m+1} \cup L> \subseteq <a_1,...,a_m,B>$. If $\rho' \in (\rho_{m+1},\rho'')$, there exists $\tilde{a}' \in \rho'$ such that $\tilde{a}' \in (a_{m+1},a'')$ and again $a_{m+1} \in <\tilde{a}',a''> \subseteq <a_1,...,a_m,B>$. Similarly if $\rho'' \in (\rho_{m+1},\rho')$ we obtain $a_{m+1} \in <a_1,...,a_m,B>$. Since $A \cup B$ is an affine basis for X it follows that, for some $i \le m$, $a_{m+1} = a_i$ and hence $\rho_{m+1} = \rho_i$.

It follows that $H_+ : L$ has finite dimension r if and only if L has codimension r in H.

It will now be shown that, whether or not a dense linear geometry has the Desargues property, its factor geometries always have the Desargues property.

PROPOSITION 6 *Let X be a dense linear geometry, H_+ an open half-space associated with a hyperplane H of X, and L a nonempty affine subset of H. Then $H_+ : L$ is a dense linear geometry with the Desargues property.*

Proof By Proposition IV.26 and Proposition 5, it only remains to prove that $H_+ : L$ has the Desargues property. If $H_+ : L$ has four affine independent points, we can apply Proposition 1. If $H_+ : L$ has at most two affine independent points, the Desargues property is vacuously satisfied. Thus we now assume that dim $H_+ : L = 2$. Then the Desargues property holds in X, by Proposition 1, and H has an affine basis of the form $D \cup \{c',c''\}$, where D is an affine basis for L. Choose $d \in D$ and $c \in (d,c'')$. Then $M = <D \setminus d,c,c'>$ is a hyperplane of H and d,c'' lie in different open half-spaces of H associated with the hyperplane M.

For any $a \in H_+$, let σ_a be the closed half-space of $<a \cup M>$ associated with the hyperplane M which contains a. We are going to show that $H_+ : M = \{\sigma_a : a \in H_+\}$ is totally ordered by writing $\sigma_a \le \sigma_b$ if $[d,b] \cap \sigma_a \ne \emptyset$.

We show first that the ordering depends only on $\sigma_b \in H_+ : M$ and not on $b \in H_+$. Since it is evident that $\sigma_a = \sigma_b$ implies $\sigma_a \le \sigma_b$, we may suppose that $\sigma_a \cap \sigma_b = M$. Assume on the contrary that $(d,b) \cap \sigma_a \ne \emptyset$, but $[d,b'] \cap \sigma_a = \emptyset$ for some $b' \in \sigma_b \setminus M$. Then d and b' lie in the same open half-space of X associated with the hyperplane $<a \cup M>$, but d and b lie in different open half-spaces. Hence b and b' lie in different open half-spaces and there exists a point $x \in (b,b') \cap \langle a \cup M \rangle$. Since $x \in \sigma_a \cap \sigma_b$, but $x \notin M$, this is a contradiction.

Suppose now that $\sigma_a \leq \sigma_b$ and $\sigma_b \leq \sigma_a$. Then there exist points $a' \in \sigma_a \backslash M$ and $b' \in \sigma_b \backslash M$ such that $a' \in [d,b]$ and $b' \in [d,a]$. Hence there exists a point $y \in [a,a'] \cap [b,b']$. Since $y \in \sigma_a \cap \sigma_b$ and $y \notin M$, it follows that $\sigma_a = \sigma_b$.

Suppose next that $\sigma_a \leq \sigma_b$ and $\sigma_b \leq \sigma_c$. Then there exists a point $b' \in \sigma_b \backslash M$ such that $b' \in [d,c]$. Also, since $\sigma_{b'} = \sigma_b$, there exists a point $a' \in \sigma_a \backslash M$ such that $a' \in [d,b']$. Then $a' \in [d,c]$ and hence $\sigma_a \leq \sigma_c$.

It remains to show that, for any $a,b \in H_+$, either $\sigma_a \leq \sigma_b$ or $\sigma_b \leq \sigma_a$. Obviously we may assume $b \notin \sigma_a$ and $a \notin \sigma_b$. Let C',C'' be the closed half-spaces of X associated with the hyperplanes $H' = \langle a \cup M \rangle$, $H'' = \langle b \cup M \rangle$ which contain d. We intend to show that either $a \in C''$ or $b \in C'$.

Assume on the contrary that $a \notin C''$ and $b \notin C'$. Then there exist points $u \in (d,a) \cap H''$ and $v \in (d,b) \cap H'$. Hence d,a,b are not collinear and there exists a point $w \in (a,v) \cap (b,u)$. Then $w \in H' \cap H'' = M$ and there exists a point $t \in (a,b)$ such that $w \in (d,t)$. Since $t \in \langle d,w \rangle \subseteq H$ and $(a,b) \subseteq H_+$, this is a contradiction.

From $c'' \in H \backslash M$ it follows that $c'' \notin H',H''$. Let B',B'' be the closed half-spaces of X associated with the hyperplanes H',H'' which contain c''. Then in the same way we can show that either $a \in B''$ or $b \in B'$. Since $b \notin \sigma_a$ we cannot have $b \in H' = C' \cap B'$, and since $a \notin \sigma_b$ we cannot have $a \in H'' = C'' \cap B''$. Hence $a \in C''$ implies $b \notin C'$, and $b \in C'$ implies $a \notin C''$. But if $b \notin C'$, then $(d,b) \cap H' \neq \emptyset$ and hence $\sigma_a \leq \sigma_b$. Similarly if $a \notin C''$, then $\sigma_b \leq \sigma_a$.

We now proceed with the proof that $H_+ : L$ has the Desargues property. Let $\langle \rho_1,\rho_1' \rangle$, $\langle \rho_2,\rho_2' \rangle$, $\langle \rho_3,\rho_3' \rangle$ be three distinct 'lines' in $H_+ : L$ with a common 'point' $\rho \neq \rho_i,\rho_i'$ $(i = 1,2,3)$. Suppose also that the intersections

$$\gamma_1 = \langle \rho_2,\rho_3 \rangle \cap \langle \rho_2',\rho_3' \rangle,\ \gamma_2 = \langle \rho_3,\rho_1 \rangle \cap \langle \rho_3',\rho_1' \rangle,\ \gamma_3 = \langle \rho_1,\rho_2 \rangle \cap \langle \rho_1',\rho_2' \rangle$$

exist. Choose $a_i \in \rho_i \backslash L$, $a_i' \in \rho_i' \backslash L$ $(i = 1,2,3)$. We begin by showing that there exists $h' \in H_+$ such that the hyperplane $H' = \langle h' \cup M \rangle$ of X intersects $(d,a_i]$, $(d,a_i']$ $(i = 1,2,3)$.

Let σ_i,σ_i' be the closed half-space of $\langle a_i \cup M \rangle$, resp. $\langle a_i' \cup M \rangle$, associated with the hyperplane M which contains a_i, resp. a_i'. By what we have just proved we may choose the notation so that $\sigma_1 \leq \sigma_i,\sigma_i'$ $(i = 1,2,3)$. Thus if we take $h' = a_1$ and $H' = \langle h' \cup M \rangle$, then $[d,a_i] \cap \sigma_1 \neq \emptyset$, $[d,a_i'] \cap \sigma_1 \neq \emptyset$ $(i = 1,2,3)$. Since

$d \notin H'$, it follows that H' contains points of $(d,a_i]$, $(d,a_i']$ $(i = 1,2,3)$. By changing notation we may now assume that $a_i, a_i' \in H'$ $(i = 1,2,3)$.

Suppose first that $L = \{d\}$. Then any element of $H_+ : L$ intersects H' in at most one point. Since $\rho \in \langle\rho_i,\rho_i'\rangle$ and $\rho \neq \rho_i,\rho_i'$ $(i = 1,2,3)$, it follows from Lemma IV.24 and the definition of a line that there exists a point $p_i \in \rho \setminus L$ such that $p_i \in \langle a_i,a_i'\rangle$ and $p_i \neq a_i,a_i'$. Moreover, by the preceding remark, $p_i = p$ is independent of i. Furthermore the lines $\langle a_i,a_i'\rangle$ are distinct, since the 'lines' $\langle\rho_i,\rho_i'\rangle$ are distinct. In the same way γ_1 contains both a point $h_1' \in \langle a_2,a_3\rangle$ and a point $h_1'' \in \langle a_2',a_3'\rangle$, and we must have $h_1' = h_1''$. Thus the intersections

$$h_1' = \langle a_2,a_3\rangle \cap \langle a_2',a_3'\rangle, \quad h_2' = \langle a_3,a_1\rangle \cap \langle a_3',a_1'\rangle, \quad h_3' = \langle a_1,a_2\rangle \cap \langle a_1',a_2'\rangle$$

exist. Hence, by the Desargues property in X, h_1',h_2',h_3' are collinear. Consequently $\gamma_1,\gamma_2,\gamma_3$ are collinear in $H_+ : L$.

Suppose next that $L \neq \{d\}$ and put $\tilde{L} = \langle D \setminus d\rangle$. Any $a \in H_+$ is contained in a unique $\rho \in H_+ : L$ and a unique $\tilde{\rho} \in H_+ : \tilde{L}$. It is easily seen that if also $a \in H'$, then $\tilde{\rho} = \rho \cap H'$. Hence if we put $\tilde{\rho}_i = \rho_i \cap H'$, $\tilde{\rho}_i' = \rho_i' \cap H'$ $(i = 1,2,3)$, then the lines $\langle\tilde{\rho}_1,\tilde{\rho}_1'\rangle, \langle\tilde{\rho}_2,\tilde{\rho}_2'\rangle, \langle\tilde{\rho}_3,\tilde{\rho}_3'\rangle$ in $H_+ : \tilde{L}$ have a common point $\neq \tilde{\rho}_i,\tilde{\rho}_i'$ $(i = 1,2,3)$. Furthermore the intersections

$$\tilde{\gamma}_1 = \langle\tilde{\rho}_2,\tilde{\rho}_3\rangle \cap \langle\tilde{\rho}_2',\tilde{\rho}_3'\rangle, \quad \tilde{\gamma}_2 = \langle\tilde{\rho}_3,\tilde{\rho}_1\rangle \cap \langle\tilde{\rho}_3',\tilde{\rho}_1'\rangle,$$
$$\tilde{\gamma}_3 = \langle\tilde{\rho}_1,\tilde{\rho}_2\rangle \cap \langle\tilde{\rho}_1',\tilde{\rho}_2'\rangle$$

exist. But the Desargues property holds in $H_+ : \tilde{L}$, since $\dim H_+ : \tilde{L} = 3$. Consequently $\tilde{\gamma}_1,\tilde{\gamma}_2,\tilde{\gamma}_3$ are collinear in $H_+ : \tilde{L}$. Since $\tilde{\gamma}_i = \gamma_i \cap H'$ $(i = 1,2,3)$, it follows that $\gamma_1,\gamma_2,\gamma_3$ are collinear in $H_+ : L$. □

5 NOTES

Among the many texts on projective geometry, two which are particularly concerned with its foundations are Heyting (1980) and the classic Veblen and Young (1910/18). A basic theorem says that if P is a projective space containing a pair of disjoint 'lines', then there is a vector space V over a division ring D such that $P = P(V)$. Moreover the same conclusion holds if no two 'lines' of P are disjoint (i.e. if P is a projective plane), provided Desargues' theorem holds in P. In the next chapter we will prove the analogues of these theorems for linear geometries, in which case the division ring D is ordered.

Desargues' theorem dates from 1636. Smith (1959) gives a short account of the life of Desargues, together with an English translation of the statement of his theorem in the *Oeuvres de Desargues*. Desargues' theorem provides an interesting combinatorial configuration of ten points and ten lines, each line containing three of the points and each point contained in three of the lines. The ubiquitous Petersen graph is obtained by taking the ten points as vertices, with two points joined by an edge if they are not both contained in one of the ten lines.

Segre (1956) has given an 'algebraic' example of a projective plane which does not have the Desargues property. Its 'points' are the 'points' (ξ_0,ξ_1,ξ_2) of the real projective plane, but its 'lines' are the two-parameter family of cubic curves

$$(p_0^2 + p_1^2 + p_2^2)(p_0\xi_0 + p_1\xi_1 + p_2\xi_2)(\xi_0^2 + \xi_1^2 + \xi_2^2) + \varepsilon p_2^3\xi_2^3 = 0,$$

where ε is a sufficiently small positive constant and p_0,p_1,p_2 are homogeneous parameters specifying the curve.

Many dense unending two-dimensional linear geometries which do not have the Desargues property are provided by Example 5, in particular the two specific cases discussed there. (Even Moulton planes with different refractive indices μ are non-isomorphic.) Rather than attempt to enumerate all such geometries, we will show in Chapter VIII that they can still possess nice 'Euclidean' properties.

The proof of Proposition 4 is based on Sperner (1938), but some extra argument is needed because we do not require that the linear geometry be unending.

VII
Vector Spaces

Here the results of the previous chapter are used to introduce coordinates, using a splendid idea of von Staudt, as modified by Hilbert and then generalized by Sperner. We define *isomorphism* of linear geometries and establish a result of Doignon, that any dense linear geometry of dimension ≥ 2 with the Desargues property can be isomorphically embedded in a *projective* space over an ordered division ring. In general such a geometry cannot be isomorphically embedded in a *vector* space over an ordered division ring, but an open half-space of such a geometry can be embedded in this way.

We also study properties involving *vector sums* of convex sets in a vector space over an ordered division ring. In particular, we establish a useful algebraic version of Rådström's cancellation law, which will be encountered in Chapter IX.

1 COORDINATES

We assume throughout this section that X is a dense linear geometry of dimension ≥ 2 *with the Desargues property*. We are going to show that it is possible to define *addition* and *multiplication* of points of a segment so that all the usual algebraic laws hold, with the exception of the commutative law for multiplication. Actually the entire discussion will be based on Proposition VI.4.

Let z,w be any two distinct points. In order to define the sum of two points $a,b \in (z,w)$, choose any point $w'' \notin <z,w>$ and any point $w' \in (w,w'')$. There exists a point $a' \in (z,w')$ such that the line $<a,a'>$ intersects the segment (w,w''). Indeed we can take any $a' \in (a'',w')$, where $a'' = (a,w'') \cap (z,w')$. Put

$$w_a = <a,a'> \cap (w,w''), \quad \tilde{p} = (b,w') \cap (a',w).$$

Then actually $w_a \in (w',w'')$ and we can define the *sum* $p = a + b$ by $p = \langle \tilde{p},w_a \rangle \cap (z,w)$. (See Figure 1.)

This definition does not depend on the choice of w' and a'. For suppose $\omega'' \notin \langle z,w \rangle$, $\omega' \in (w,\omega'')$ and $\alpha' \in (z,\omega')$ is such that $\langle a,\alpha' \rangle$ intersects (w,ω'') in ω_a. Define $\tilde{\pi}$ and π in the analogous way to $\tilde{p}$ and p. Applying Proposition VI.4 to the quadruples $w',\tilde{p},a',w_a$ and $\omega',\tilde{\pi},\alpha',\omega_a$, we immediately obtain $\pi = p$.

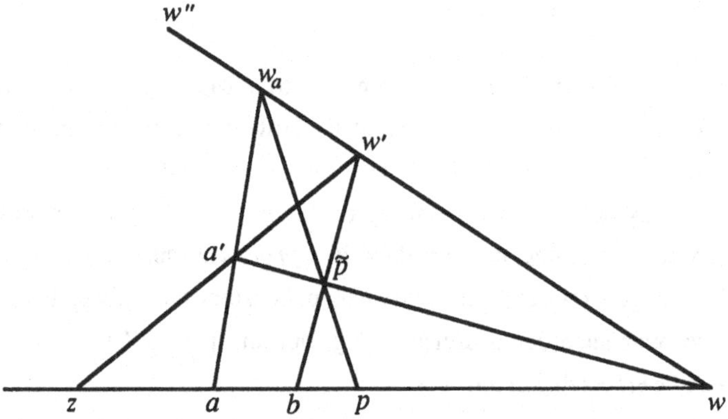

Figure 1: Addition

We show first that addition is *commutative*: $a + b = b + a$. Evidently we may assume that $a \neq b$. Put $b' = (b,w_a) \cap (z,w')$ and $\tilde{\rho} = (a,w') \cap (b',w)$. Then (with b' in the role of a') $\rho = b + a$ is given by $\rho = \langle \tilde{\rho},w_a \rangle \cap (z,w)$. Applying Proposition VI.4 to the quadruples $a',\tilde{p},w',w_a$ and $w',w_a,b',\tilde{\rho}$, we immediately obtain $\rho = p$. (See Figure 2.)

We show next that addition is *associative*: $(a + b) + c = a + (b + c)$. Besides the notations already introduced, put $\tilde{w} = \langle z,\tilde{p} \rangle \cap (w,w'')$ and $\tilde{q} = (c,\tilde{w}) \cap (a',w)$. Then (with $\tilde{w},\tilde{p}$ in the role of w',a') $q = b + c$ is given by $q = \langle \tilde{q},w' \rangle \cap (z,w)$, and similarly $s = p + c$ is given by $s = \langle \tilde{q},w_a \rangle \cap (z,w)$. Since (with w',a' as initially) $r = a + q$ is also given by $r = \langle \tilde{q},w_a \rangle \cap (z,w)$, we obtain $r = s$. (See Figure 3.)

It is easily seen that, for any points $a,b \in (z,w)$, there exists a point $c \in (z,w)$ such that $b = a + c$ if and only if $a \in (z,b)$. Moreover, c is then uniquely determined and $c \in (z,b)$.

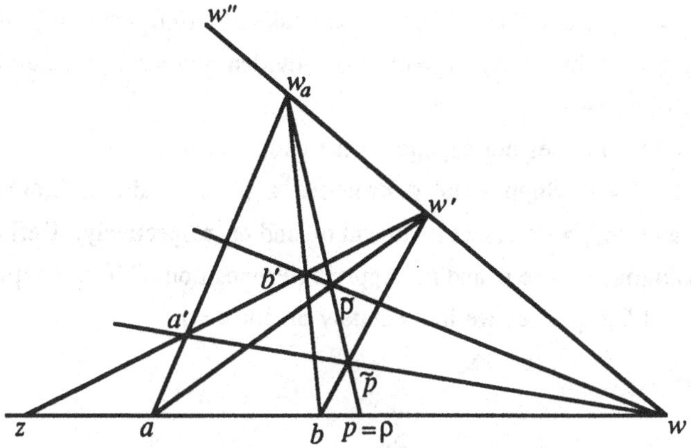

Figure 2: Commutativity of addition

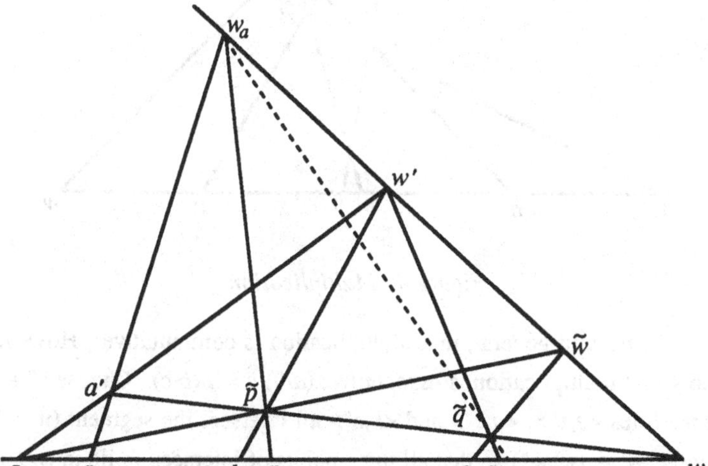

Figure 3: Associativity of addition

To define the product of two points $a,b \in (z,w)$, we fix a point $e \in (z,w)$. There exists a point $z' \in (z,w')$ such that the lines $\langle e,z' \rangle$ and $\langle a,z' \rangle$ both intersect the segment (w,w''). Indeed we can take any $z' \in (z'',w')$, where $z'' = (e,w'') \cap (z,w')$ if $a \in (z,e]$ and $z'' = (a,w'') \cap (z,w')$ if $a \in [e,w)$. Put

$$w_e = \langle e,z'\rangle \cap (w,w''), \quad w_a = \langle a,z'\rangle \cap (w,w'').$$

Then actually $w_e, w_a \in (w',w'')$. Thus we can take $p' = (b,w_e) \cap (z,w')$ and define the *product* $p = a \cdot b$ by $p = \langle p',w_a\rangle \cap (z,w)$. Evidently $p = b$ if $a = e$ and $p = a$ if $b = e$. (See Figure 4.)

This definition does not depend on the choice of w' and z'. To see this, we may assume $a,b \neq e$. Suppose $\omega'' \notin \langle z,w\rangle$, $\omega' \in (w,\omega'')$ and $\zeta' \in (z,\omega')$ are such that $\langle e,\zeta'\rangle$ and $\langle a,\zeta'\rangle$ intersect (w,ω'') in ω_e and ω_a respectively. Define π' and π in the analogous way to p' and p. Applying Proposition VI.4 to the quadruples z',w_a,w_e,p' and $\zeta',\omega_a,\omega_e,\pi'$, we immediately obtain $\pi = p$.

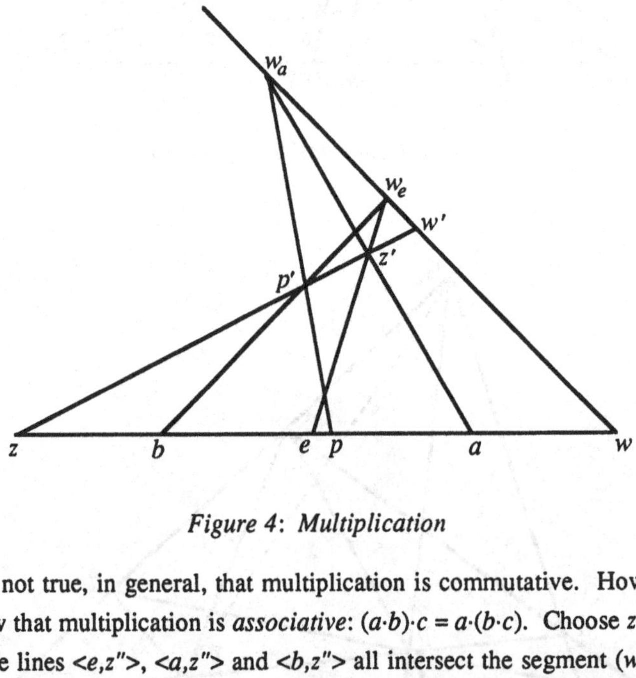

Figure 4: Multiplication

It is not true, in general, that multiplication is commutative. However, we now show that multiplication is *associative*: $(a \cdot b) \cdot c = a \cdot (b \cdot c)$. Choose $z'' \in (z,w')$ so that the lines $\langle e,z''\rangle$, $\langle a,z''\rangle$ and $\langle b,z''\rangle$ all intersect the segment (w',w'') and then choose $w_e \in (w',w'')$ so that all three points of intersection lie in (w_e,w''). If we put $z' = (e,w_e) \cap (z,w')$, then actually $z' \in (z'',w')$. Hence there exist points $w_a = \langle a,z'\rangle \cap (w',w'')$ and $p' = (b,w_e) \cap (z,w')$. Furthermore, there exist points $w_p = \langle e,p'\rangle \cap (w',w'')$ and $q' = (c,w_p) \cap (z,w')$. Then $p = a \cdot b$ is given by $p = \langle p',w_a\rangle \cap (z,w)$ and (with p' in the role of z') $q = b \cdot c$ is given by $q = \langle q',w_e\rangle \cap (z,w)$. Since $q' = (c,w_p) \cap (z,w')$, $s = p \cdot c$ is given by $s = \langle q',w_a\rangle \cap (z,w)$, and since $q' = (q,w_e) \cap (z,w')$, $r = a \cdot q$ is given by $r = \langle q',w_a\rangle \cap (z,w)$. Thus $r = s$.

It will now be shown that the points of the segment (z,w) form a *group* under multiplication. Since multiplication is associative and e acts as an identity element, we need only show that for each point $a \in (z,w)$ there exists a point $b \in (z,w)$ such that $a \cdot b = e$. But if $e' = (e,w_a) \cap (z,w')$ and $b = <e',w_e> \cap (z,w)$, then $a \cdot b = e$.

We prove next the *distributive* laws: $(a + b) \cdot c = a \cdot c + b \cdot c$ and $c \cdot (a + b) = c \cdot a + c \cdot b$. Again put $p = a + b$ and choose $z' \in (z,w')$ so that the lines $<e,z'>$, $<a,z'>,<b,z'>,<p,z'>$ all intersect the segment (w,w''), in points w_e,w_a,w_b,w_p say. If $q' = (c,w_e) \cap (z,w')$ then, by the definition of multiplication,

$$a \cdot c = <q',w_a> \cap (z,w), \quad b \cdot c = <q',w_b> \cap (z,w), \quad p \cdot c = <q',w_p> \cap (z,w).$$

By the definition of addition $p = <\tilde{p},w_a> \cap (z,w)$, where $\tilde{p} = (b,w') \cap (z',w)$. If $s' = (q',w) \cap (b.c,w')$ then, by applying Proposition VI.4 to the quadruples $b,p,z',\tilde{p}$ and $b \cdot c,p \cdot c,q',s'$, we obtain $w_a \in <p \cdot c,s'>$. Since $w_a \in <a \cdot c,q'>$, this proves that $p \cdot c = a \cdot c + b \cdot c$.

To prove the second distributive law choose $z' \in (z,w')$ so that the lines $<e,z'>$ and $<c,z'>$ intersect the segment (w',w''), in w_e and w_c. Then

$$c \cdot a = <a',w_c> \cap (z,w), \quad c \cdot b = <b',w_c> \cap (z,w),$$

where

$$a' = (a,w_e) \cap (z,w'), \quad b' = (b,w_e) \cap (z,w').$$

Moreover $p = a + b$ and $r = c \cdot p$ are given by

$$p = <\tilde{p},w_e> \cap (z,w), \quad \tilde{p} = (a',w) \cap (b,w'),$$

and

$$r = <p',w_c> \cap (z,w), \quad p' = (p,w_e) \cap (z,w').$$

If we put $\tilde{r} = (a',w) \cap (r,w_c)$ then, by applying Proposition VI.4 to the quadruples $w,b,w_e,\tilde{p}$ and $w,c \cdot b,w_c,\tilde{r}$, we obtain $w' \in <c \cdot b,\tilde{r}>$. This proves that $r = c \cdot a + c \cdot b$.

Put $P = (z,w)$ and let D denote the collection of all ordered pairs (a_1,a_2) with $a_1,a_2 \in P$. Two elements (a_1,a_2) and (a_1',a_2') of D will be said to be *equal* if there exist $x,x' \in P$ such that

$$a_1 + x = a_1' + x', \quad a_2 + x = a_2' + x'.$$

It is easily verified that this is indeed an equivalence relation on D. Moreover, for any $x \in P$, $(a_1{+}x,a_2{+}x) = (a_1,a_2)$. We also define the sum and product of two elements of D by

$$(a_1,a_2) + (b_1,b_2) = (a_1+b_1,a_2+b_2), \quad (a_1,a_2)\cdot(b_1,b_2) = (a_1\cdot b_1+a_2\cdot b_2,a_1\cdot b_2+a_2\cdot b_1).$$

It is easily verified that these definitions do not depend on the choice of elements within an equivalence class, i.e., if $(a_1',a_2') = (a_1,a_2)$ and $(b_1',b_2') = (b_1,b_2)$, then

$$(a_1',a_2') + (b_1',b_2') = (a_1,a_2) + (b_1,b_2), \quad (a_1',a_2')\cdot(b_1',b_2') = (a_1,a_2)\cdot(b_1,b_2).$$

Furthermore, the commutative and associative laws for addition, the associative law for multiplication and the two distributive laws hold in D, since they hold in P.

The elements $0 = (a,a)$ and $1 = (e+a,a)$ act as identity elements for addition and multiplication respectively. Moreover any element (a_1,a_2) of D has an additive inverse (a_2,a_1). We will show that any element $(a_1,a_2) \neq 0$ has a multiplicative inverse. Since $a_1 \neq a_2$, either $a_1 \in (z,a_2)$ or $a_2 \in (z,a_1)$. If $a_1 \in (z,a_2)$, there exists $x \in P$ such that $a_1 + x = a_2$ and hence $(a_1,a_2)\cdot(a,x^{-1}+a) = 1$. If $a_2 \in (z,a_1)$, there exists $y \in P$ such that $a_2 + y = a_1$ and hence $(a_1,a_2)\cdot(y^{-1}+a,a) = 1$.

Thus we have now shown that D is a division ring. Since the map $x \to (x+a,a)$ of P into D preserves sums and products, we may identify the point $x \in P$ with the element $(x+a,a)$ of D. In this way P is embedded in D. Since P is closed under addition and multiplication, and D is the disjoint union of the sets P, $\{0\}$ and $-P$, it follows that D is an *ordered division ring* with P as the subset of *positive* elements.

We now show that if a and b are distinct points of P, then the 'open segment' (a,b) is the set of all points $c \in P$ which can be represented in the form $c = \lambda a + (1-\lambda)b$, where $0 < \lambda < 1$. Without loss of generality assume $a < b$, i.e., $a \in (z,b)$, and suppose first that c admits such a representation. Since $\lambda a < \lambda b$ we have $c < \lambda b + (1-\lambda)b = b$, and since $(1-\lambda)a < (1-\lambda)b$ we have $a = \lambda a + (1-\lambda)a < c$. Thus $c \in (z,b)$ and $a \in (z,c)$, which implies $c \in (a,b)$.

Suppose on the other hand that $c \in (a,b)$. Since $a \in (z,c)$ we have $c = a + \alpha$, where $\alpha \in P$, and since $c \in (z,b)$ we have $b = c + \gamma$, where $\gamma \in P$. Thus $b = a + \beta$, where $\beta = \alpha + \gamma$ and hence $\gamma < \beta$. If we put $\lambda = \gamma\beta^{-1}$, then $0 < \lambda < 1$ and

$$c = a + \alpha = a + (1-\lambda)\beta = \lambda a + (1-\lambda)b.$$

2 ISOMORPHISMS

Two linear geometries X and Y are said to be *isomorphic* if there is a one-to-one map f of X onto Y such that $f([x_1,x_2]) = [f(x_1),f(x_2)]$ for all $x_1,x_2 \in X$. The map f itself is said to be an *isomorphism* of X onto Y. It is easily seen that f^{-1} is an isomorphism of Y onto X. Moreover, if g is an isomorphism of Y onto Z, then $g \circ f$ is an isomorphism of X onto Z. If ℓ is a line in X then $f(\ell)$ is a line in Y, and if A is an affine independent subset of X then $f(A)$ is an affine independent subset of Y.

With this terminology, we have shown in Section 1 that the segment (z,w) of a dense linear geometry X of dimension ≥ 2 with the Desargues property is isomorphic to the set P of positive elements of an ordered division ring D. The segment (z,w) of X is also isomorphic to the segment $(0,1)$ of D, since the map $f: x \rightarrow x(1+x)^{-1}$ of P onto $(0,1)$ is an isomorphism. It follows that the segment $[z,w]$ of X is isomorphic to the segment $[0,1]$ of D.

PROPOSITION 1 *Let X be a dense linear geometry of dimension ≥ 2. If $a_1,a_2,b_1,b_2 \in X$ and $a_1 \neq a_2$, $b_1 \neq b_2$, then the segments $[a_1,a_2]$ and $[b_1,b_2]$ are isomorphic.*

Proof Suppose first that $b_1 = a_1$ and $b_2 \notin \langle a_1,a_2\rangle$. Choose some $c \in (a_2,b_2)$. For any $a \in (a_1,a_2)$, there exist a unique $a' \in (b_2,a) \cap (b_1,c)$ and a unique $b \in (b_1,b_2)$ such that $a' \in (a_2,b)$. Moreover every $b \in (b_1,b_2)$ may be obtained in this way. If we put $f(a) = b, f(a_1) = b_1, f(a_2) = b_2$, then it follows from Propositions II.17′ and II.2′ that f is an isomorphism of $[a_1,a_2]$ onto $[b_1,b_2]$.

Suppose next that $\langle a_1,a_2\rangle = \langle b_1,b_2\rangle$ and choose $c_2 \notin \langle a_1,a_2\rangle$. By what we have already proved, $[a_1,a_2]$ is isomorphic to $[a_1,c_2]$, $[a_1,c_2]$ is isomorphic to $[b_1,c_2]$, and $[b_1,c_2]$ is isomorphic to $[b_1,b_2]$. Hence $[a_1,a_2]$ is isomorphic to $[b_1,b_2]$.

Suppose finally that $\langle a_1,a_2\rangle \neq \langle b_1,b_2\rangle$. We may choose the notation so that $b_1 \notin \langle a_1,a_2\rangle$ and $a_1 \notin \langle b_1,b_2\rangle$. Then, by what we have already proved, $[a_1,a_2]$ is isomorphic to $[a_1,b_1]$ and $[a_1,b_1]$ is isomorphic to $[b_1,b_2]$. Hence again $[a_1,a_2]$ is isomorphic to $[b_1,b_2]$. □

That the hypothesis on the dimension cannot be omitted in Proposition 1 is shown, for example, by the dense totally ordered set X consisting of all rational numbers in the interval $[0,1]$ and all real numbers outside this interval.

PROPOSITION 2 *Let X and $\tilde{X}$ be dense linear geometries with the Desargues property. Let A_0 be an affine independent subset of X containing at least two points and let f_0 be an isomorphism of $C_0 = [A_0]$ onto $\tilde{C}_0 \subseteq \tilde{X}$. If $a_1 \in X \setminus <A_0>$ and $\tilde{a}_1 \in \tilde{X} \setminus <\tilde{A}_0>$, where $\tilde{A}_0 = f_0(A_0)$, then there exists an isomorphism f of $C = [A_0 \cup a_1]$ onto $\tilde{C} = [\tilde{A}_0 \cup \tilde{a}_1]$ such that $f(a_1) = \tilde{a}_1$ and $f(x) = f_0(x)$ for $x \in C_0$.*

Proof For convenience of writing put $A = A_0 \cup a_1$ and $\tilde{A} = \tilde{A}_0 \cup \tilde{a}_1$. Also, put $C_1 = [A \setminus a_0]$ where $a_0 \in A_0$, and $\tilde{C}_1 = [\tilde{A} \setminus \tilde{a}_0]$ where $\tilde{a}_0 = f_0(a_0)$. Further, put $A_0' = A_0 \setminus a_0$, $C_0' = [A_0']$ and $\tilde{A}_0' = \tilde{A}_0 \setminus \tilde{a}_0$, $\tilde{C}_0' = [\tilde{A}_0']$. Assume initially that there exist $b \in X$ such that $a_1 \in (a_0,b)$ and $\tilde{b} \in \tilde{X}$ such that $\tilde{a}_1 \in (\tilde{a}_0, \tilde{b})$.

For any $x_1 \in C_1 \setminus C_0'$ with $x_1 \neq a_1$, there are a unique point $x_1' \in C_0'$ such that $x_1 \in (a_1,x_1')$ and a unique point $x_0 \in (a_0,x_1')$ such that $x_1 \in (b,x_0)$. If we put $\pi(x_1) = x_0$, $\pi(a_1) = a_0$ and $\pi(x) = x$ for $x \in C_0'$, then π maps C_1 isomorphically onto C_0. In the same way, with b replaced by $\tilde{b}$, we can define an isomorphism $\tilde{\pi}$ of $\tilde{C}_1$ onto $\tilde{C}_0$ with $\tilde{\pi}(\tilde{a}_1) = \tilde{a}_0$ and $\tilde{\pi}(y) = y$ for $y \in \tilde{C}_0'$. Then $f_1 = \tilde{\pi}^{-1} \circ f_0 \circ \pi$ is an isomorphism of C_1 onto $\tilde{C}_1$ such that $f_1(a_1) = \tilde{a}_1$ and $f_1(x) = f_0(x)$ for $x \in C_0' = C_0 \cap C_1$.

For any $x \in C \setminus [a_0,a_1]$, there exist a unique $x_0 \in C_0 \setminus a_0$ such that $x \in (a_1,x_0]$ and a unique $x_1 \in C_1 \setminus a_1$ such that $x \in (a_0,x_1]$. Moreover, there exist a unique $x_0' \in C_0'$ such that $x_0 \in (a_0,x_0']$ and a unique $x_1' \in C_0'$ such that $x_1 \in (a_1,x_1']$. Then $x_0',x_1' \in <a_0,a_1,x>$, since $x_0,x_1 \in <a_0,a_1,x>$. Since also $x_0',x_1' \in <A_0'>$, $x_0' \neq x_1'$ would imply $<a_0,x_0',x_1'> = <a_0,a_1,x>$ and hence $a_1 \in <A_0>$, which is a contradiction. Consequently $x_0' = x_1' = x'$, say.

It follows that $f_0(x_0) \in (\tilde{a}_0, \tilde{x}']$ and $f_1(x_1) \in (\tilde{a}_1, \tilde{x}']$, where $\tilde{x}' = f_0(x') = f_1(x')$. Moreover, if $x \notin C_0 \cup C_1$, then $f_0(x_0),f_1(x_1) \neq \tilde{x}'$ and hence there exists a unique point $\tilde{x} \in (\tilde{a}_0,f_1(x_1)) \cap (\tilde{a}_1,f_0(x_0))$. We define a map f: $C \setminus (a_0,a_1) \to \tilde{C}$ by putting $f(x) = \tilde{x}$ if $x \notin C_0 \cup C_1$, $f(x) = f_0(x)$ if $x \in C_0$ and $f(x) = f_1(x)$ if $x \in C_1$.

To define $f(z)$ for $z \in (a_0,a_1)$ choose $a \in A_0'$ and $x \in (a,z)$. Then $f(x)$ is in the intrinsic interior of $[\tilde{a}_0, \tilde{a}_1,f(a)]$ and hence $f(x) \in (f(a), \tilde{z})$ for some $\tilde{z} \in (\tilde{a}_0, \tilde{a}_1)$. We wish to show that $\tilde{z}$ is independent of the choice of a and x. This is obvious if we only change x. If $a' \in A_0' \setminus a$ and $x' \in (a',z)$, then $f(a),f(x),f(x'),f(a')$ are coplanar and hence $\tilde{z}' = \tilde{z}$. We now complete the definition of f by putting $f(z) = \tilde{z}$. It is not difficult to verify that f is an isomorphism of C onto $\tilde{C}$ with the required properties.

It remains to show that our initial assumption may be removed. Choose $a_0' \in A_0'$ and then $c_0 \in (a_0, a_0')$ and $a_1' \in (c_0, a_1)$. Also, let $\tilde{a}_0' = f_0(a_0')$, $\tilde{c}_0 = f_0(c_0)$, and choose $\tilde{a}_1' \in (\tilde{c}_0, \tilde{a}_1)$. Then $a_1' \in (a_0, b)$ for some $b \in (a_0', a_1)$ and $\tilde{a}_1' \in (\tilde{a}_0, \tilde{b})$ for some $\tilde{b} \in (\tilde{a}_0', \tilde{a}_1)$. By what we have already proved, there exists an isomorphism f'' of $C' = [A_0 \cup a_1']$ onto $\tilde{C}' = [\tilde{A}_0 \cup \tilde{a}_1']$ such that $f''(a_1') = \tilde{a}_1'$ and $f''(x) = f_0(x)$ for $x \in C_0$. Clearly it is enough now to show that there exists an isomorphism ϕ of C onto C' such that $\phi(a_1) = a_1'$ and $\phi(x) = x$ for $x \in C_0$. We will merely give the definition of ϕ at other points.

For any $x \in C \setminus C_0$ with $x \neq a_1$, there is a unique point $x_0 \in C_0$ such that $x \in (a_1, x_0)$. If $x_0 \neq c_0$ we put $\phi(x) = x'$, where $x' = (a_1', x_0) \cap (x, c_0)$. If $x \in (a_1, c_0)$, there is a unique $x'' \in (a_1, a_0)$ such that $x \in (x'', a_0')$ and we define $\phi(x) = (\phi(x''), a_0') \cap (a_1', c_0)$. □

A subset of a linear geometry is said to be a *simplex* if it is the convex hull of an affine independent set. Our labours so far in this chapter are essentially summarised in the following characterization of simplices:

THEOREM 3 *Let X be a dense linear geometry with* $\dim X > 2$, *or with* $\dim X = 2$ *and the Desargues property. Then there is an ordered division ring D such that any simplex $C \subseteq X$ is isomorphic to a simplex $\tilde{C}$ in a vector space $\tilde{X}$ over D.*

Proof We have already shown that there exists an ordered division ring D such that, for any affine independent set $A \subseteq X$ with $|A| = 2$, there is an affine independent set $\tilde{A} \subseteq D$ with $|\tilde{A}| = 2$ such that $[A]$ is isomorphic to $[\tilde{A}]$.

Suppose now that A is an affine independent set with $|A| > 2$. Let $\mathcal{F}$ be the family of all triples $(B, \tilde{X}, f)$, where $B \subseteq A$, $\tilde{X}$ is a vector space over D and f is an isomorphism of $[B]$ onto a simplex in $\tilde{X}$. We partially order $\mathcal{F}$ by writing $(B, \tilde{X}, f) \preccurlyeq (B', \tilde{X}', g)$ if $B \subseteq B'$, $\tilde{X}$ is a subspace of $\tilde{X}'$ and $g(x) = f(x)$ for $x \in [B]$.

Let $\mathcal{T} = \{(B_i, \tilde{X}_i, f_i)\}$ be a maximal totally ordered subfamily of $\mathcal{F}$ and put $B = \bigcup B_i$, $\tilde{X} = \bigcup \tilde{X}_i$. If $x \in [B]$, then $x \in [B_i]$ for some i and we can define $f: [B] \to \tilde{X}$ by $f(x) = f_i(x)$. Moreover $f(B)$ is an affine independent set, since any finite subset is affine independent, and hence f is an isomorphism of $[B]$ onto the simplex $[f(B)] = \bigcup [f(B_i)]$. Thus $(B, \tilde{X}, f)$ is an element of $\mathcal{F}$ and $(B_i, \tilde{X}_i, f_i) \preccurlyeq (B, \tilde{X}, f)$ for all i. If $B = A$ we are finished. If not, then there exists a point $a_1 \in A \setminus B$. If $\tilde{X}'$ is a vector space over the division ring D which properly contains $\tilde{X}$, then, by Proposition 2, there exists an element $(B \cup a_1, \tilde{X}', g)$ of $\mathcal{F}$ such that $(B, \tilde{X}, f) \preccurlyeq (B \cup a_1, \tilde{X}', g)$. But this contradicts the maximality of $\mathcal{T}$. □

In Theorem 3 the whole space X need not be isomorphic to a convex subset of a vector space over an ordered division ring (even if X is also unending). This is shown by Example VI.6, since any two coplanar lines of a projective space intersect. However, the whole space X may be characterized in the following way:

THEOREM 4 *Let X be a dense linear geometry with* $\dim X > 2$, *or with* $\dim X = 2$ *and the Desargues property. Then there exists a vector space V over an ordered division ring D such that X is isomorphic to a set X' in the projective completion $\overline{V}$ of V where, for any $y',z' \in X'$ with $y' \neq z'$, $[y',z']$ is one of the two closed segments with endpoints y',z' of the 'line' $\langle y',z' \rangle$.*

Proof Let $E = \{e_i : i \in I\}$ be an affine basis for X. By Theorem 3 there exists an isomorphism f of the simplex $C = [E]$ onto a simplex $C' = [E']$ in a vector space V over an ordered division ring D. Moreover we may assume that $E' = \{e_i' = f(e_i) : i \in I\}$ is an affine basis for V. Let $\overline{V}$ be the projective completion of V.

If $x \in X \setminus C$, then there exist lines through x which contain more than one point of C. For let $\{e_1,\dots,e_n\}$ be a finite subset of E, with $n \geq 3$, such that $x \in \langle e_1,\dots,e_n \rangle$ and put $P = [e_1,\dots,e_n]$. For any $p \in P^i$ the line $\langle p,x \rangle$ intersects P in a nondegenerate segment, since $x \notin P$.

For any $x \in X$ let α,β be distinct lines through x, each containing more than one point of C, and let α', β' be the images under f of $\alpha \cap C$, $\beta \cap C$. The distinct coplanar lines $\langle \alpha' \rangle$, $\langle \beta' \rangle$ intersect in a unique point x' of $\overline{V}$. We are going to show that x' does not depend on the choice of lines α,β. Clearly it is sufficient to show that if γ is a third line through x containing more than one point of C, and γ' the image under f of $\gamma \cap C$, then $x' \in \langle \gamma' \rangle$.

Let a_1,a_2 be points of $\alpha \cap C$ with $a_2 \in (x,a_1)$, let b_1,b_2 be points of $\beta \cap C$ with $b_2 \in (x,b_1)$, and let c_1,c_2 be points of $\gamma \cap C$ with $c_1 \in (x,c_2)$. Choose $d_1 \in (a_1,b_1)$ and then $d_2 \in (x,d_1)$ so that $d_2 \in (a_2,w)$ for some $w \in (d_1,b_1)$ and $d_2 \in (b_2,\tilde{w})$ for some $\tilde{w} \in (a_1,d_1)$. Then $d_1,d_2,w,\tilde{w} \in C$ and

$$w = \langle a_1,d_1 \rangle \cap \langle a_2,d_2 \rangle, \quad \tilde{w} = \langle b_1,d_1 \rangle \cap \langle b_2,d_2 \rangle.$$

Let δ' be the image under f of $\delta \cap C$, where $\delta = \langle d_1,d_2 \rangle$. Since the intersections

$$u = (a_1,c_1) \cap (a_2,c_2), \quad v = (c_1,d_1) \cap (c_2,d_2)$$

exist and are points of C, and since the lines $\langle a_1,a_2 \rangle, \langle c_1,c_2 \rangle, \langle d_1,d_2 \rangle$ have the common point x, it follows from the Desargues property that u,v,w are collinear (see Figure 5). Under f the points $a_i,\ldots,d_i$ are mapped into points $a_i',\ldots,d_i'$ of C'

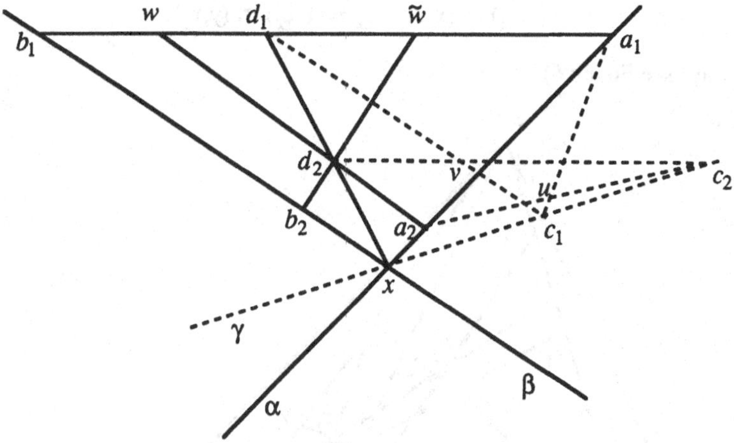

Figure 5

with preservation of segments and collinearity. Thus u',v',w' are collinear points of $\overline{V}$. Since $\overline{V}$ has the converse Desargues property (Proposition VI.3), it follows that the lines $\langle \alpha' \rangle, \langle \gamma' \rangle, \langle \delta' \rangle$ have a common point of intersection. Since the intersection $\tilde{u} = (b_1,c_1) \cap (b_2,c_2)$ also exists and is a point of C, it follows in the same way that the lines $\langle \beta' \rangle, \langle \gamma' \rangle, \langle \delta' \rangle$ have a common point of intersection. Hence the lines $\langle \alpha' \rangle, \langle \beta' \rangle, \langle \gamma' \rangle$ have a common point of intersection.

We now extend the definition of f to X by putting $f(x) = x'$. We are going to show next that distinct collinear points of X are mapped by f into distinct collinear points of $\overline{V}$. This implies, in particular, that f is injective.

Let x,y,z be points of X such that $x \in (y,z)$ and let x',y',z' be their images under f. There exists a finite subset $\{e_1,\ldots,e_n\}$ of E, with $n \geq 3$, such that $y,z \in \langle e_1,\ldots,e_n \rangle$. Put $P = [e_1,\ldots,e_n]$ and choose $p \in P^i$ with $p \notin \langle y,z \rangle$. There exist points $\tilde{x}, \bar{y}, \bar{z} \in C$ such that $p \in (x,\tilde{x})$, $\bar{y} \in (p,y)$, $\bar{z} \in (p,z)$. There exist also points $\bar{x} = (p,x) \cap (\bar{y},\bar{z})$, $y_1 \in (\tilde{x},\bar{z})$ such that $p \in (\bar{y},y_1)$, and $z_1 \in (\tilde{x},\bar{y})$ such that $p \in (\bar{z},z_1)$. Choose $\tilde{y} \in (\bar{y},z_1)$ and let $x_1 \in (\tilde{x},y_1)$ be such that $\tilde{y} \in (y,x_1)$. If we put $a_1 = p$, then there exist points

$$b_1 = (\bar{z},a_1) \cap (x,x_1), \quad c_1 = (b_1,x_1) \cap (a_1,y_1).$$

Let $a_2 \in (z,c_1)$ be such that $b_1 \in (y,a_2)$, let $y_2 \in (\bar{z},y_1)$ be such that $a_2 \in (b_1,y_2)$, and let $z_2 \in (\tilde{x},z_1)$ be such that $c_1 \in (a_2,z_2)$. Finally, if we choose $x_2 \in (\tilde{x},z_2)$, there exist points

$$b_2 = (z,z_2) \cap (x,x_2), \quad c_2 = (x,x_2) \cap (y,y_2).$$

In this way (see Figure 6)

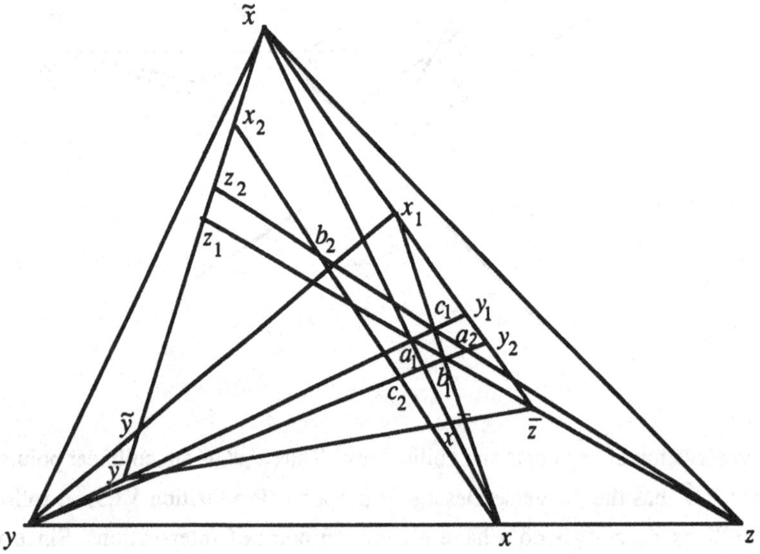

Figure 6

$$a_i = (y,y_i) \cap (z,z_i), \quad b_i = (z,z_i) \cap (x,x_i), \quad c_i = (x,x_i) \cap (y,y_i) \quad (i = 1,2)$$

and

$$(b_1,b_2) \cap (c_1,c_2) \neq \emptyset, \quad (c_1,c_2) \cap (a_1,a_2) \neq \emptyset, \quad (a_1,a_2) \cap (b_1,b_2) \neq \emptyset.$$

Since a_1,b_1,c_1 and a_2,b_2,c_2 are non-collinear triples, and since

$$x = \langle b_1,c_1 \rangle \cap \langle b_2,c_2 \rangle, \quad y = \langle a_1,c_1 \rangle \cap \langle a_2,c_2 \rangle, \quad z = \langle a_1,b_1 \rangle \cap \langle a_2,b_2 \rangle$$

are distinct collinear points, it follows from Proposition VI.3 that the segments (a_1,a_2), (b_1,b_2), (c_1,c_2) have a common point q. Since $\bar{x}, \bar{y}, \bar{z} \in C$, it follows that

$$x' = \langle b_1',c_1'\rangle \cap \langle b_2',c_2'\rangle,\ y' = \langle a_1',c_1'\rangle \cap \langle a_2',c_2'\rangle,\ z' = \langle a_1',b_1'\rangle \cap \langle a_2',b_2'\rangle$$

and $\langle a_1',a_2'\rangle$, $\langle b_1',b_2'\rangle$, $\langle c_1',c_2'\rangle$ have the common point $q' \neq a_i',b_i',c_i'$ $(i = 1,2)$. Hence x',y',z' are distinct and collinear, by the Desargues property in $\bar{V}$.

It will now be shown that f maps the segment $[y,z]$ of X onto the segment of the line $\langle y',z'\rangle$ in $\bar{V}$ which has endpoints y',z' and contains x'. Suppose that $w \in (y,z)$, where $w \neq x$, and put $\bar{w} = (p,w) \cap (\bar{y}, \bar{z})$. Then the lines $\langle y',\bar{y}'\rangle$, $\langle z',\bar{z}'\rangle, \langle x',\bar{x}'\rangle, \langle w',\bar{w}'\rangle$ have the common point p' and $\bar{x}', \bar{w}' \in \langle \bar{y}', \bar{z}'\rangle$, $x',w' \in \langle y',z'\rangle$. Since $\bar{x}', \bar{w}'$ lie in the same open segment with endpoints $\bar{y}', \bar{z}'$, it follows that x',w' lie in the same open segment with endpoints y',z'.

On the other hand, if w' is any point of the open segment with endpoints y',z' which contains x', the point $\bar{w}' = \langle p',w'\rangle \cap \langle \bar{y}', \bar{z}'\rangle$ lies in the open segment with endpoints $\bar{y}', \bar{z}'$ which contains $\bar{x}'$. Hence $\bar{w}' = f(\bar{w})$ for some $\bar{w} \in (\bar{y}, \bar{z})$ and there exists a point $w \in (y,z)$ such that $\bar{w} \in (p,w)$. It follows that $w' = f(w)$. Thus f maps X isomorphically onto $X' = f(X)$. □

Although in Theorem 4 the whole space X need not be isomorphic to a convex set in a vector space over an ordered division ring, it will now be shown that any open half-space of X is isomorphic to such a convex set.

THEOREM 5 *Let X be a dense linear geometry with* $\dim X > 2$, *or with* $\dim X = 2$ *and the Desargues property. If H_+ is an open half-space of X associated with a hyperplane H, then H_+ is isomorphic to a convex subset of a vector space V over an ordered division ring D.*

Proof Let $E = \{e_i : i \in I\}$ be an affine basis for X such that $H = \langle E \setminus e_1\rangle$ and $e_1 \in H_+$. By Theorem 3 there exists an isomorphism f of the simplex $C = [E]$ onto a simplex $C' = [E']$ in a vector space V over an ordered division ring D. Moreover we may assume that $V = \langle E'\rangle$. By Theorem 4 and its proof we may extend f to an isomorphism of X into the projective completion $\bar{V}$ of V. The hyperplane $\langle E' \setminus e_1'\rangle$ of V extends to a hyperplane H' of $\bar{V}$. We will show that $f(H_+) \cap H' = \emptyset$.

Assume on the contrary that $x' := f(x) \in H'$ for some $x \in H_+$. Let $\{e_1,\ldots,e_n\}$ be a finite subset of E, with $n \geq 3$, such that $x \in \langle e_1,\ldots,e_n\rangle$. Put $P = [e_1,\ldots,e_n]$

and $Q = [e_2,...,e_n]$. If $h \in Q^i$ then $<x,h> \cap P = [h,h_+]$, where $h_+ \in H_+$. Thus there exist points $y,z \in (h,x) \cap C$ with $y \neq z$. If h',y',z' are the images of h,y,z under f, then h',y',z' are collinear and so are x',y',z'. Since $y',z' \notin H'$, it follows that $h' = x'$. Since this holds for any $h \in Q^i$ and f is injective on C, we have a contradiction.

By definition, $\overline{V} = P(V^*)$, where V^* is a vector space over the division ring D. There exists a bijective linear map $\gamma: V^* \to V^*$ inducing a bijective map $g: \overline{V} \to \overline{V}$ such that $g(H')$ is the hyperplane at infinity of V. Then $g \circ f$ maps H_+ isomorphically onto a convex subset of V. □

In the statement of Theorem 5 we cannot replace 'open half-space' by 'closed half-space'. This is shown by Example VI.7. In fact, since any two coplanar lines of the projective space $X = P^n$ $(n \geq 2)$ intersect, X is not isomorphic to a convex subset of a vector space over an ordered division ring. However, X is itself the closed half-space associated with the hyperplane P^{n-1}.

3 VECTOR SUMS

Throughout this section we assume that X is a vector space over an ordered division ring D with the standard alignment of convex sets.

For any nonempty sets $A,B \subseteq X$ and any $\alpha \in D$, we define

$$A + B = \{a + b: a \in A, b \in B\},$$
$$\alpha A = \{\alpha a: a \in A\}.$$

The set $A + B$ will be called the *vector* (or *Minkowski*) *sum* of A and B, and αA will be called a *scalar multiple* of A.

It is easily verified that the following properties hold for any nonempty sets $A,B,C \subseteq X$ and any $\alpha,\beta \in D$:

(i) $A + B = B + A$, $(A + B) + C = A + (B + C)$,
(ii) $\alpha(A + B) = \alpha A + \alpha B$, $(\alpha\beta)A = \alpha(\beta A)$, $1A = A$.

The singleton $0 = \{0\}$ acts as a zero element:

(iii) $A + 0 = A$, $0A = 0$.

Furthermore,

(iv) *if* $A \subseteq B$, *then* $A + C \subseteq B + C$ *and* $\alpha A \subseteq \alpha B$.

We consider now the case where the sets are convex.

LEMMA 6 *If* A,B *are nonempty convex subsets of* X *and* $\alpha \in D$, *then* $A + B$ *and* αA *are also nonempty convex subsets of* X. *Moreover*

(v) $(\lambda + \mu)A = \lambda A + \mu A$ *for any* $\lambda,\mu \in D$ *with* $\lambda,\mu \geq 0$.

Proof Since A,B are convex, for any $\lambda,\mu \geq 0$ with $\lambda + \mu = 1$ we have $\lambda A + \mu A \subseteq A$ and $\lambda B + \mu B \subseteq B$. Hence

$$\lambda(A + B) + \mu(A + B) = (\lambda A + \mu A) + (\lambda B + \mu B) \subseteq A + B$$

and

$$\lambda(\alpha A) + \mu(\alpha A) = \alpha(\alpha^{-1}\lambda\alpha A + \alpha^{-1}\mu\alpha A) \subseteq \alpha A.$$

Thus $A + B$ and αA are convex.

For any nonempty set $A \subseteq X$ and any $\lambda,\mu \in D$ we have $(\lambda + \mu)A \subseteq \lambda A + \mu A$. Hence to prove the last statement of the lemma we need only show that if A is convex and if $\lambda,\mu \geq 0$, $\lambda + \mu > 0$, then $\lambda A + \mu A \subseteq (\lambda + \mu)A$. But if $x = \lambda a' + \mu a''$, where $a',a'' \in A$, then

$$x = (\lambda + \mu)\{(\lambda + \mu)^{-1}\lambda a' + (\lambda + \mu)^{-1}\mu a''\} \in (\lambda + \mu)A. \quad \square$$

It may be noted that A is necessarily convex if $(\lambda + \mu)A = \lambda A + \mu A$ for all $\lambda,\mu \geq 0$, and that if $|A| > 1$, $\lambda > 0$ then

$$0 = (\lambda + (-\lambda))A \neq \lambda A + (-\lambda)A.$$

It follows by induction from Lemma 6 that, for any nonempty convex set A and any positive integer n,

$$nA = A + \ldots + A,$$

where there are n summands on the right.

PROPOSITION 7 *If* $S,T \subseteq X$, *then* $[S + T] = [S] + [T]$.

Proof If $x \in [S + T]$, then there exist $s_j \in S$ and $t_j \in T$ $(j = 1,\ldots,n)$ such that

$$x = \sum_{j=1}^{n} \lambda_j(s_j + t_j),$$

where $\lambda_j \geq 0$ $(j = 1,\ldots,n)$ and $\sum_{j=1}^{n} \lambda_j = 1$. Hence $x = y + z$, where $y = \sum_{j=1}^{n} \lambda_j s_j \in [S]$ and $z = \sum_{j=1}^{n} \lambda_j t_j \in [T]$. Thus $[S + T] \subseteq [S] + [T]$.

On the other hand, if $x \in [S] + [T]$, then there exist $s_i \in S$ $(i = 1,\ldots,m)$ and $t_j \in T$ $(j = 1,\ldots,n)$ such that

$$x = \sum_{i=1}^{m} \lambda_i s_i + \sum_{j=1}^{n} \mu_j t_j,$$

where $\lambda_i \geq 0$ $(i = 1,\ldots,m)$, $\mu_j \geq 0$ $(j = 1,\ldots,n)$ and $\sum_{i=1}^{m} \lambda_i = 1 = \sum_{j=1}^{n} \mu_j$. Hence

$$x = \sum_{i=1}^{m} \sum_{j=1}^{n} \lambda_i \mu_j (s_i + t_j)$$

and $x \in [S + T]$. Thus $[S] + [T] \subseteq [S + T]$. □

PROPOSITION 8 *If A and B are convex sets such that $A^i \neq \emptyset$ and $B^i \neq \emptyset$, then*

$$(A + B)^i = A^i + B^i.$$

Proof We show first that $A^i + B^i \subseteq (A + B)^i$. Let $z = x + y$, where $x \in A^i$ and $y \in B^i$, and let $c = a + b$, where $a \in A$, $b \in B$ and $c \neq z$. There exists $\mu > 1$ such that, for $1 < \lambda < \mu$,

$$x' = \lambda x + (1 - \lambda)a \in A, \quad y' = \lambda y + (1 - \lambda)b \in B.$$

Then, for $1 < \lambda < \mu$,

$$\lambda z + (1 - \lambda)c = x' + y' \in A + B$$

and hence $z \in (A + B)^i$.

We show next that $(A + B)^i \subseteq A^i + B^i$. Let $x \in A^i$ and $y \in B^i$. Then $z = x + y \in (A + B)^i$ by what we have just proved. Suppose $w \in (A + B)^i$ and $w \neq z$. Since $z \in A + B$, we have $w \in (z,c)$ for some $c \in A + B$. Thus $c = a + b$, where $a \in A$ and $b \in B$, and, for some θ such that $0 < \theta < 1$, $w = \theta z + (1 - \theta)c$. Hence $w = x' + y'$, where $x' = \theta x + (1 - \theta)a \in A^i$ and $y' = \theta y + (1 - \theta)b \in B^i$. Thus $w \in A^i + B^i$. □

PROPOSITION 9 *If C_1 and C_2 are nonempty convex sets, and if F is a nonempty face of $C = C_1 + C_2$, then there exist nonempty faces F_1 of C_1 and F_2 of C_2 such that $F = F_1 + F_2$.*

Proof Put

$$F_1 = \{x \in C_1 : x + y \in F \text{ for some } y \in C_2\},$$
$$F_2 = \{y \in C_2 : x + y \in F \text{ for some } x \in C_1\}.$$

Evidently F_1 and F_2 are nonempty convex sets.

Suppose $x_1 \in F_1$ and $x_1 = \theta c_1 + (1-\theta)c_1'$, where $c_1, c_1' \in C_1$ and $0 < \theta < 1$. Then $x_1 + y \in F$ for some $y \in C_2$. Since $x_1 + y = \theta(c_1 + y) + (1-\theta)(c_1' + y)$ and F is a face of C, it follows that $c_1, c_1' \in F_1$. Hence F_1 is a face of C_1, and similarly F_2 is a face of C_2.

Obviously $F \subseteq F_1 + F_2$. On the other hand, if $x_1 \in F_1$ and $y_2 \in F_2$, then there exist $y_1 \in C_2$ and $x_2 \in C_1$ such that $x_1 + y_1 \in F$ and $x_2 + y_2 \in F$. Hence

$$(1/2)(x_1 + y_2) + (1/2)(x_2 + y_1) = (1/2)(x_1 + y_1) + (1/2)(x_2 + y_2) \in F.$$

Since F is a face of C, it follows that $x_1 + y_2 \in F$. Thus $F_1 + F_2 \subseteq F$. □

PROPOSITION 10 *Let $C = A + B$, where A and B are nonempty convex sets. If z is an extreme point of C, then it has a unique representation in the form $z = x + y$, where $x \in A$ and $y \in B$. Moreover, in this representation x is an extreme point of A and y an extreme point of B.*

Proof By Proposition 9 there exists a representation $z = x + y$, where x is an extreme point of A and y an extreme point of B. Suppose also that $z = x' + y'$, where $x' \in A$ and $y' \in B$. Since $z = (1/2)(x + y') + (1/2)(x' + y)$ and z is an extreme point of C, we must have $x + y' = x' + y$. Then $2z = 2x + y + y' = 2x' + y + y'$ and hence $x = x'$, $y = y'$. □

PROPOSITION 11 *Let A, C, P be nonempty sets, with A arbitrary, C convex and P a polytope. If $A + P \subseteq C + P$, then $A \subseteq C$.*

Proof We wish to show that if $a + P \subseteq C + P$, then $a \in C$. Let $P = [p_1,\dots,p_n]$ be the convex hull of n points. Since the result is obvious if $n = 1$, we assume that $n > 1$ and the result holds for polytopes which are the convex hulls of fewer than n points. Hence if we put $Q = [p_2,\dots,p_n]$, it is sufficient to show that $a + Q \subseteq C + Q$.

If $q \in Q$, then $a + q = c + p$ for some $c \in C$ and $p \in P$. Similarly $a + p_1 = c' + p'$ for some $c' \in C$ and $p' \in P$. But $P = [p_1 \cup Q]$. Hence if $p \notin Q$, then $p = \alpha p_1 + (1-\alpha)q'$, where $q' \in Q$ and $0 < \alpha \le 1$. Similarly if $a \ne c'$, then $p' = \beta p_1 + (1-\beta)q''$, where $q'' \in Q$ and $0 \le \beta < 1$. It follows that

$$p_1 = q'' + (1-\beta)^{-1}(c' - a)$$

and hence

$$a + q = c + \alpha q'' + \alpha(1-\beta)^{-1}(c' - a) + (1-\alpha)q'.$$

Consequently

$$[1 + \alpha(1-\beta)^{-1}](a+q) = c + \alpha(1-\beta)^{-1}c' + \alpha q'' + (1-\alpha)q' + \alpha(1-\beta)^{-1}q.$$

Since C and Q are convex, this shows that $a + q \in C + Q$. Thus $a + Q \subseteq C + Q$. □

PROPOSITION 12 *If P_1 and P_2 are polytopes, then $P = P_1 + P_2$ is also a polytope. Moreover any nonempty face F of P has a unique representation $F = F_1 + F_2$, where F_k is a nonempty face of P_k $(k = 1,2)$.*

Proof It follows at once from Proposition 7 that P is a polytope and from Proposition 9 that any nonempty face F of P has a representation $F = F_1 + F_2$, where F_1 is a nonempty face of P_1 and F_2 a nonempty face of P_2. We suppose F_1 and F_2 defined as in the proof of Proposition 9.

Suppose $F = F_1' + F_2'$, where F_1' is a nonempty face of P_1 and F_2' a nonempty face of P_2. Then $F_k' \subseteq F_k$ $(k = 1,2)$, by the definition of F_1, F_2 in the proof of Proposition 9. Since

$$F = F_1' + F_2' \subseteq F_1' + F_2 \subseteq F_1 + F_2 = F,$$

it follows that $F_1' + F_2 = F = F_1 + F_2$. Hence, by Proposition 11, $F_1' = F_1$ and similarly $F_2' = F_2$. □

PROPOSITION 13 *If P is a polytope and A,B nonempty convex sets such that $A + B = P$, then A and B are also polytopes.*

Proof Let $P = [p_1,\ldots,p_n]$. Then $p_k = a_k + b_k$ for some $a_k \in A$ and $b_k \in B$ $(k = 1,\ldots,n)$. If $A' = [a_1,\ldots,a_n]$ and $B' = [b_1,\ldots,b_n]$, then $A' \subseteq A$, $B' \subseteq B$ and hence $A' + B' \subseteq P$. On the other hand, $p_k \in A' + B'$ $(k = 1,\ldots,n)$ and $A' + B'$ is convex, by Lemma 6. Hence $P \subseteq A' + B'$. Thus $P = A' + B'$, and so $A + B' \subseteq A' + B'$. Hence $A \subseteq A'$, by Proposition 11. Consequently $A = A'$, and similarly $B = B'$. □

4 NOTES

It was first shown by von Staudt (1856/57) that it is possible to introduce coordinates in space, independently of the notions of distance and congruence. His discussion was based on the 'fundamental theorem of projective geometry', which may be given the following formulation:

If a map of a line to itself is the product of finitely many perspectivities and has at least three fixed points, then it is the identity map.

Some interesting remarks on the fundamental theorem of projective geometry are contained in Pickert (1981) and Frank (1992).

Hilbert (1899) showed that the Desargues property could be used instead of the fundamental theorem of projective geometry. In a noteworthy (but rather inaccessible) paper, Sperner (1938) showed that Hilbert's argument could be adapted to any unending two-dimensional linear geometry with the Desargues property. Doignon (1976), using ideas of Busemann (1955), extended Sperner's work to higher dimensions and proved Theorems 3–5 for any unending linear geometry. The more general treatment here for dense, rather than unending, linear geometries presents little difficulty after the corresponding extension in Proposition VI.4.

The division ring in these theorems is a field, i.e. multiplication is commutative, if and only if the fundamental theorem of projective geometry holds or, equivalently, if and only if the linear geometry has the following 'Pappus property':

Suppose a_1,b_1,c_1 are distinct collinear points, and a_2,b_2,c_2 are distinct collinear points not on the same line; if $\langle a_1,b_2\rangle$ intersects $\langle a_2,b_1\rangle$, $\langle b_1,c_2\rangle$ intersects $\langle b_2,c_1\rangle$, and $\langle c_1,a_2\rangle$ intersects $\langle c_2,a_1\rangle$, then the three points of intersection are collinear.

Furthermore the Pappus property actually implies the Desargues property. (It is pointed out in Seidenberg (1976) that the original proof of this by Hessenberg (1905) was incomplete.) Proofs of these results for ordinary projective space are given in Heyting (1980), for example. We do not establish them here, since commutativity of multiplication will be a consequence of the additional axiom **(S)** in Chapter VIII.

Proposition 10 appeared in Klee (1959). Proposition 11 is a topology-free version of Rådström's cancellation law (Proposition IX.4). It may be noted that Proposition 11 no longer holds if 'polytope' is replaced by 'polyhedron'. For example, if $X = \mathbb{R}^2$, $A = \{(1,1)\}$ and $C = \{(0,0)\}$ are singletons, and $P = \{(x,y) \in \mathbb{R}^2: x,y \geq 0\}$ is the first quadrant, then

$$A + P \subset P = C + P.$$

Bair and Fourneau (1975/80) contains additional results on vector sums, and additional references.

VIII
Completeness

In this chapter we introduce a final axiom of *completeness*, which is an extension to linear geometries of the Dedekind cut property for the real line. Indeed it is shown that, in any complete dense linear geometry of dimension ≥ 2, a segment containing more than one point is isomorphic to the interval $[0,1]$ of $\mathbb{R}$. It is further shown that, in any complete dense unending linear geometry, two convex sets with disjoint nonempty intrinsic interiors can be *properly separated* by a hyperplane. This in turn implies the geometric form of the Hahn–Banach theorem and a separation theorem for any finite number of convex sets. These results hold even if the linear geometry is two-dimensional and does not have the Desargues property.

We prove next the *fundamental theorem of ordered geometry* that any complete unending linear geometry of dimension ≥ 2 with the Desargues property is isomorphic to a convex set, which is its own intrinsic interior, in a real vector space. The fundamental theorem may be established in other ways than that used here, but they have the disadvantage of requiring dimension ≥ 3. The necessity of the various hypotheses of the theorem is illustrated by examples. It is somewhat surprising that a finite-dimensional projective space over any ordered division ring other than $\mathbb{R}$ can be given the structure of a dense unending linear geometry. At the end of the chapter we recall, for use in Chapter IX, some well-known properties of metric spaces and normed vector spaces.

1 INTRODUCTION

A linear geometry X will be said to be *complete* if the following axiom is satisfied:

(**S**) *if* $a,b \in X$ *and if* C *is a convex subset of* $[a,b]$ *such that* $a \in C$, *then there exists a point* $c \in [a,b]$ *such that either* $C = [a,c]$ *or* $C = [a,c)$.

The axiom's label is chosen to indicate that c 'separates' the points of $[a,b]$ which are in C from those which are not. It is easily seen that the point c is uniquely determined if the linear geometry X is dense. On the other hand, if $[a,b] = \{a,b\}$ for some distinct $a,b \in X$ and if $C = \{a\}$, then we can take either $c = a$ or $c = b$.

The axiom (S) may be regarded as an extension to linear geometries of the Dedekind cut property for the real line $\mathbb{R}$. Any interval of $\mathbb{R}$ is a complete dense linear geometry. However, a segment in a complete dense linear geometry need not be isomorphic to an interval of $\mathbb{R}$:

EXAMPLE 1 Let W be an uncountable set, e.g. the set of all sequences (a_n), where $a_n \in \{0,1\}$. By the well-ordering principle, there is a total ordering of W such that every nonempty subset of W has a least element. (The well-ordering principle may be deduced from Hausdorff's maximality theorem; see, e.g., Hewitt and Stromberg (1975).) Let Z be the set of all $z \in W$ for which the set $\{w \in W: w < z\}$ is uncountable. If Z is nonempty, it has a least element ζ. Put $Y = \{y \in W: y < \zeta\}$ if $Z \neq \emptyset$ and $Y = W$ otherwise. Then Y is an uncountable totally ordered set with the following properties:

(i) every nonempty subset of Y has a least element,
(ii) for each $y \in Y$, the set $\{y' \in Y: y' < y\}$ is countable,
(iii) Y has no greatest element.

The *long line* Λ is defined to be the set of all pairs (y,ρ), where $y \in Y$ and $\rho \in [0,1)$, except for the pair $(y_0,0)$ where y_0 is the least element of Y, with the lexicographic ordering defined by $(y_1,\rho_1) \prec (y_2,\rho_2)$ if $y_1 < y_2$, or if $y_1 = y_2$ and $\rho_1 < \rho_2$. Since Λ is a dense totally ordered set with no least or greatest element, it is a dense unending linear geometry with the definition of segments given in Example VI.1. Furthermore the separation axiom (S) is satisfied. Since Λ has no countable dense subset, it is not isomorphic to $\mathbb{R}$. However, any segment of Λ containing more than one point is isomorphic to the interval $[0,1]$ of $\mathbb{R}$.

Finally define a totally ordered set $X = \bigcup_{i=1}^{3} X_i$, where X_1 and X_3 are copies of the long line Λ, $X_2 = \{p\}$ and $x_j < x_k$ if $x_j \in X_j, x_k \in X_k$ $(j < k)$. Then X is a complete dense unending linear geometry, but the segment $[a,b]$ is not isomorphic to an interval of $\mathbb{R}$ if $a \in X_1$ and $b \in X \setminus X_1$.

In Example 1 all points of X are collinear. It will now be shown that if X is a complete dense linear geometry, not all points of which are collinear, then any segment of X *is* isomorphic to an interval of $\mathbb{R}$. We first prove

LEMMA 1 *Let S be a countable dense totally ordered set with no least or greatest element. Then S is order isomorphic to the set* $\mathbb{Q}$ *of rational numbers.*

Proof Let $\{s_1,s_2,\ldots\}$ and $\{r_1,r_2,\ldots\}$ be enumerations of S and $\mathbb{Q}$ respectively. We wish to construct a bijective map $\varphi: S \to \mathbb{Q}$ such that $\varphi(s_j) \gtreqless \varphi(s_k)$ according as $s_j \gtreqless s_k$ for all distinct $j,k \in \mathbb{N}$. The construction is inductive.

Put $\varphi(s_1) = r_1$ and write $s^1 = s_1$, $r^1 = r_1$. Suppose we have defined subsets $S^n = \{s^1,\ldots,s^n\}$, $\mathbb{Q}^n = \{r^1,\ldots,r^n\}$ of S, $\mathbb{Q}$ respectively and a bijective map $\varphi: S^n \to \mathbb{Q}^n$ such that $\varphi(s^j) \gtreqless \varphi(s^k)$ according as $s^j \gtreqless s^k$ for all disinct $j,k \in \{1,\ldots,n\}$. If n is odd define r^{n+1} to be the rational number r_k with least k which is not in $\mathbb{Q}^n$. Then either r^{n+1} is less than all elements of $\mathbb{Q}^n$, or r^{n+1} is greater than all elements of $\mathbb{Q}^n$, or there exist two elements r^j,r^k of $\mathbb{Q}^n$ such that $r^j < r^{n+1} < r^k$, but $r^j < r^i < r^k$ for no $r^i \in \mathbb{Q}^n$. We define s^{n+1} to be the element s_k of S with least k which is correspondingly less than all elements of S^n, or greater than all elements of S^n, or satisfies $s^j < s^{n+1} < s^k$, and we put $\varphi(s^{n+1}) = r^{n+1}$. If n is even we interchange the roles of S and $\mathbb{Q}$. Thus we define s^{n+1} to be the element s_k of S with least k which is not in S^n and we define $r^{n+1} = \varphi(s^{n+1})$ to be the rational number r_k with least k which has the same order relation with respect to $\mathbb{Q}^n$ as s^{n+1} has with respect to S^n. Since S and $\mathbb{Q}$ are both dense, and have no least or greatest element, this construction is always possible. Moreover each element of S and $\mathbb{Q}$ is ultimately included. □

PROPOSITION 2 *If X is a complete dense linear geometry with* $\dim X \geq 2$, *then every nondegenerate segment of X is isomorphic to the interval* $[0,1]$ *of* $\mathbb{R}$.

Proof Let a and b be distinct points of X and choose $y_0 \in (a,b)$. Define a sequence $\{y_m\}$ recurrently by $y_m \in (a,y_{m-1})$ $(m \geq 1)$. If we suppose the segment $[a,b]$ totally ordered in accordance with Proposition III.4 so that $a < b$, then $\{y_m\}$ is a decreasing sequence. Denote by C the set of all points $x \in [a,b]$ such that $x < y_m$ for all m. Then C is convex, $a \in C$ and $b \notin C$. Hence, by (S), there exists a point

$a' \in [a,b)$ such that $[a,a') \subseteq C$ and $(a',b] \subseteq X \setminus C$. By Proposition VII.1 it is enough to show that the segment $[a',b]$ is isomorphic to an interval of the real line. Thus we now take $a = a'$. Then for each $x \in (a,b)$ there is a positive integer m such that $y_m \in (a,x)$. By Proposition III.4 we need only show that there is a bijective map of (a,b) onto $\mathbb{R}$ which preserves order.

We show first that, for any $y \in (a,b)$, there exists an isomorphism f_y of $[a,b]$ onto $[y,b]$ such that $f_y(a) = y$ and $f_y(b) = b$. Choose $d \notin \langle a,b \rangle$ and $c \in (b,d)$. For any $x \in (a,b)$, there exists a unique $x' \in (a,c) \cap (d,x)$. If we put $f_1(x) = x'$, $f_1(a) = a, f_1(b) = c$, then f_1 is an isomorphism of $[a,b]$ onto $[a,c]$. For any $x' \in (a,c)$, there exists a unique $y' \in (c,y) \cap (b,x')$. If we put $f_2(x') = y', f_2(a) = y$, $f_2(c) = c$, then f_2 is an isomorphism of $[a,c]$ onto $[y,c]$. For any $y' \in (c,y)$, there exists a unique $z \in (y,b)$ such that $y' \in (d,z)$. If we put $f_3(y') = z, f_3(y) = y$, $f_3(c) = b$, then f_3 is an isomorphism of $[y,c]$ onto $[y,b]$. Evidently $f_y = f_3 \circ f_2 \circ f_1$ is an isomorphism of $[a,b]$ onto $[y,b]$ such that $f_y(a) = y$ and $f_y(b) = b$. (In the notation of Chapter VII, $f_y(x) = x + y$.) Moreover $f_y(x) \in (x,b)$ for any $x \in (a,b)$, since $y' \in (x',b)$, and for any $z \in (x,b)$ there exists $y \in (a,z)$ such that $f_y(x) = z$ (see Figure 1). Furthermore $f_y(x) < f_y(\bar{x})$ if $x < \bar{x}$, and $f_y(x) < f_{\bar{y}}(x)$ if $y < \bar{y}$.

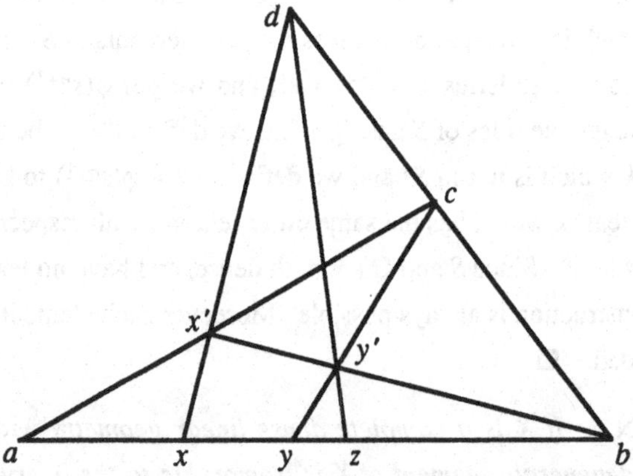

Figure 1: $z = f_y(x)$

For any $y \in (a,b)$, the iterates $y^{(0)} = y, y^{(1)} = f_y(y), y^{(2)} = f_y(f_y(y)),...$ form an increasing sequence. We now show that, for each $x \in (a,b)$, there exists a non-negative integer n such that $y^{(n)} > x$. Assume on the contrary that $y^{(n)} \leq x$ for all $n \geq 0$. It follows from (S) that there exists $z \in (a,b)$ such that $y^{(n)} \leq z$ for all $n \geq 0$, but $y^{(n)} > z'$ for some $n \geq 0$ if $z' < z$. Since $z \in (y,b)$, there exists $w \in (a,b)$ such that $f_y(w) = z$. Then $w < z$ and hence $y^{(n)} > w$ for some $n \geq 0$. Consequently $y^{(n+1)} > f_y(w) = z$, which is a contradiction.

We show next that the set E of all points $y_m^{(n)}$, where $\{y_m\}$ is the sequence constructed at the outset and n is any non-negative integer, is dense in $[a,b]$. Suppose $u,v \in (a,b)$ and $u < v$. Then there exists $t \in (a,v)$ such that $f_t(u) = v$. Choose m so that $y_m < t$ and $y_m < u$. There is an integer $p \geq 0$ such that $y_m^{(p)} \leq u < y_m^{(p+1)}$. Then $y_m^{(p+1)} \leq f_y(u) < f_t(u) = v$. Thus $y_m^{(p+1)} \in (u,v)$, as we wished to prove.

Thus (a,b) is a totally ordered set with no least or greatest element and with a countable dense subset E. Consequently, by Lemma 1, there is a bijective order-preserving map φ of E onto the set $\mathbb{Q}$ of rational numbers. For any $c \in (a,b) \setminus E$, put

$$A = \{x \in E: x < c\}, \quad B = \{x \in E: x > c\}.$$

Then A,B are disjoint nonempty subsets of E and $E = A \cup B$. Hence $\mathbb{Q}$ is the union of the disjoint nonempty subsets $\varphi(A),\varphi(B)$, and $\alpha < \beta$ for every $\alpha \in \varphi(A)$, $\beta \in \varphi(B)$. By the Dedekind property for $\mathbb{R}$, there is a unique $\gamma \in \mathbb{R} \setminus \mathbb{Q}$ such that $\alpha < \gamma < \beta$ for every $\alpha \in \varphi(A)$, $\beta \in \varphi(B)$. It is easily seen that if we extend the definition of φ to the whole of (a,b) by putting $\varphi(c) = \gamma$, then φ is a bijective order-preserving map of (a,b) onto $\mathbb{R}$. □

Suppose X is the open unit disc in the Euclidean plane, with the usual definition of segments. Through any point not on a given line there pass infinitely many lines which do not intersect the given line. However, there are two 'special' ones, namely the intersections with X of the lines in $\mathbb{R}^2$ which pass through the two endpoints on the unit circle of the given line. This example may be borne in mind in the statement and proof of the following proposition.

PROPOSITION 3 *Suppose X is a complete unending linear geometry. Let a_1,a_2,b_1 be non-collinear points of X and let H_+ be the open half-plane of the plane $\Pi = \langle a_1,a_2,b_1 \rangle$ associated with the line $H = \langle a_1,b_1 \rangle$ which contains a_2.*

Then there exists $b_2 \in H_+$ such that, for any $x \in H_+$, $[a_1,a_2\rangle \cap [b_1,x\rangle \neq \emptyset$ if and only if x lies in the open half-plane of Π associated with the line $\langle b_1,b_2 \rangle$ which contains a_1.

Furthermore $\langle a_1,a_2\rangle \cap \langle b_1,b_2\rangle = \emptyset$ *and, if the lines* $\langle a_1,a_2\rangle,\langle b_1,b_2\rangle$ *are ordered so that* $a_1 < a_2$, $b_1 < b_2$, *then for any* $a',a'' \in \langle a_1,a_2\rangle$ *with* $a' < a''$ *and any* $b',b'' \in \langle b_1,b_2\rangle$ *with* $b' < b''$, b'' *has the same properties with respect to* a',a'',b' *as* b_2 *has with respect to* a_1,a_2,b_1.

Proof Choose w so that $b_1 \in (a_1,w)$. Then, for any $x \in H_+$, the ray $[b_1,x\rangle$ contains a unique point $c \in (a_1,a_2) \cap [a_2,w)$. We wish to show that there exists $b_2 \in (a_2,w)$ such that $[b_1,c\rangle$ intersects $[a_1,a_2\rangle$ if and only if $c \in [a_2,b_2)$.

Let C be the set of all $c \in [a_2,w)$ such that $[b_1,c\rangle$ intersects $[a_1,a_2\rangle$. Evidently $a_2 \in C$ and C is convex. If we choose a_3 so that $a_2 \in (a_1,a_3)$, then there exists a point $u \in (b_1,a_3) \cap (a_2,w)$ and $u \in C$. If we choose a_0 so that $a_1 \in (a_0,a_2)$, then $b_1 \in (a_0,v)$ for some $v \in (a_2,w)$ and $v \notin C$. It now follows from (S) that there exists $b_2 \in (a_2,w)$ such that $C = [a_2,b_2]$ or $C = [a_2,b_2)$. It remains to show that $b_2 \notin C$.

Assume on the contrary that there exists a point $p \in [a_1,a_2\rangle \cap [b_1,b_2\rangle$. Then $b_2 \in (b_1,p)$ and $a_2 \in (a_1,p)$. If we choose a_3 so that $p \in (a_2,a_3)$, then there exist a point $q \in (a_3,b_1) \cap (w,p)$ and a point $r \in (b_1,q) \cap (w,b_2)$. Hence $r \in C \cap (\Pi \setminus C)$, which is a contradiction.

Suppose the lines $\ell = \langle a_1,a_2\rangle$, $m = \langle b_1,b_2\rangle$ ordered so that $a_1 < a_2$, $b_1 < b_2$. Since $b_2 \in H_+$, we have both $[a_1,a_2\rangle \cap [b_1,b_2\rangle = \emptyset$ and $\ell \cap [b_1,b_2\rangle = \emptyset$. We will show that actually $\ell \cap m = \emptyset$.

Assume on the contrary that there exists a point $p \in \ell \cap m$. Then $p < b_1$ on m, and hence $p < a_1$ on ℓ. If we choose $a_0 \in \ell$ so that $a_0 < p$, then $b_1 \in (a_0,q)$ for some $q \in (a_2,b_2)$. Hence there exists a point $r \in \ell \cap [b_1,q\rangle$. Since $b_1 \in (a_0,r)$ and $b_1 \notin \ell$, this is a contradiction.

Let a',a'' be points of ℓ with $a' < a''$ and let b',b'' be points of m with $b' < b''$. It is evident that if $a' = a_1$ and $b' = b_1$, then b'' has the same properties with respect to the ordered triple a',a'',b' as b_2 has with respect to the ordered triple a_1,a_2,b_1. The same conclusion holds if $b' = b_1$ and $a_1 < a'$, since $[a',a''\rangle \cap [b_1,x\rangle = \emptyset$ for any $x \in H_-$ in the open half-plane of Π associated with the line $\langle a',b_1\rangle$ which contains a''. The same conclusion holds also if $a' = a_1$ and $b_1 < b'$, since $[a_1,a''\rangle \cap [b',x\rangle \neq \emptyset$ for any $x \in H_-$ in the open half-plane of Π associated with the line $\langle a_1,b'\rangle$ which contains a''. Since $\ell \cap m = \emptyset$, it follows from what has already been proved and the total ordering of the lines ℓ,m that the same conclusion holds also in the general case. □

The uniquely determined ray $[b_1,b_2>$ will be called the *asymptote* through b_1 to the ray $[a_1,a_2>$. The second part of Proposition 3 shows that it depends only on the oriented line $<a_1,a_2>$, and not on the choice of a_1,a_2. It further shows that, in a sense, it depends only on the oriented line $<b_1,b_2>$.

In the same way, reversing the orientation on $<a_1,a_2>$, there exists an asymptote $[b_1,b_2{}^{\dagger}>$ through b_1 to the ray $[a_2,a_1>$. In Euclidean space $<b_1,b_2> = <b_1,b_2{}^{\dagger}>$, but the restriction of the Euclidean plane to the open unit disc already shows that this need not be true in general.

Proposition 3 shows that, for any line ℓ and any point $p \notin \ell$, there is at least one line m in the plane $<p \cup \ell>$ such that $p \in m$ and $\ell \cap m = \emptyset$. Thus we may regard Proposition 3 as a replacement for Euclid's parallel axiom. However, unlike parallelism in Euclidean space, the asymptote relationship may be neither symmetric nor transitive. To see this, in the second specific case of Example VI.5 take $a_1 = (-1,1)$, $a_2 = (-1,2)$ and $b_1 = (0,1)$, $b_2 = (0,2)$. Then $<a_1,a_2> \cap <b_1,b_2> = \emptyset$ since, in the notation of Chapter VI, $<a_1,a_2> = h^{-1}$ and $<b_1,b_2> = h^0$. Furthermore, if $a \in (a_1,a_2>$ then $[a,b_1]$ is the usual straight line segment, since $a = (-1,\alpha)$ for some $\alpha > 1$ and hence $<a,b_1> = \ell^{1-\alpha,1}$. It follows that $[b_1,b_2>$ is the asymptote through b_1 to the ray $[a_1,a_2>$. However, if $a_3 = (-1/2,2)$, it is easily seen that the asymptote through a_1 to $[b_1,b_2>$ is the ray $[a_1,a_3>$ of g.

PROPOSITION 4 *If X is a complete unending linear geometry with* $\dim X \geq 2$, *then every line of X is isomorphic to the real line* $\mathbb{R}$.

Proof It follows from Proposition 2 that, for any distinct points $a,b \in X$, (a,b) is order isomorphic to $(0,1)$ and hence to $\mathbb{R}$. We need only show that an arbitrary line $<a_1,a_2>$ of X is *order* isomorphic to $\mathbb{R}$.

Let b_1 be a point not on the line $<a_1,a_2>$, choose w so that $b_1 \in (a_1,w)$, and let $b_2 \in (a_2,w)$ be defined as in the proof of Proposition 3. Also, choose a',a'' so that $a_2 \in (a_1,a')$ and $a_1 \in (a_2,a'')$. By the proof of Proposition 3, $(a_2,a'> = [a_2,a'> \setminus a_2$ is order isomorphic to (a_2,b_2), and hence to $\mathbb{R}$ and to $(1,\infty)$. Similarly $(a_1,a''>$ is order isomorphic to $(a_1,b_2{}^{\dagger})$, and hence to $\mathbb{R}$ and to $(-\infty,0)$. Since $[a_1,a_2]$ is order isomorphic to $[0,1]$, the result follows. □

2 SEPARATION PROPERTIES

Throughout this section we will assume that a set X is given on which a complete dense unending linear geometry is defined. We draw attention to the fact that we do not assume the Desargues property. Although the Desargues property necessarily holds if $\dim X > 2$, the results which will be established are also valid if $\dim X = 2$ and the Desargues property does not hold.

The notion of 'convex partition' was introduced in Chapter II. It will now be shown that, under the present hypotheses, convex partitions are 'generally' determined by a hyperplane.

PROPOSITION 5 *If C, D is a convex partition of X such that $C^i \cup D^i \neq \emptyset$, then C^i, D^i are both nonempty and $H: = X \setminus (C^i \cup D^i)$ is a hyperplane. Moreover, the two open half-spaces associated with H are C^i, D^i and the two closed half-spaces associated with H are $\overline{C}, \overline{D}$.*

Proof By Proposition V.28, $C^i \neq \emptyset$ and $D^i \neq \emptyset$. It follows from Proposition V.28 also that $\overline{C} = X \setminus D^i$, $\overline{D} = X \setminus C^i$ and hence $H = \overline{C} \cap \overline{D}$. Thus H is convex. In fact H is affine. For suppose $a,b \in H$ and $a \in (b,c)$. Then $c \notin C^i$, since $b \in \overline{C}$ and $a \notin C^i$. Similarly $c \notin D^i$, and thus $c \in H$.

Let $c \in C^i$ and $d \in D^i$. By (S), the segment $[c,d]$ contains a point x such that $[c,x) \subseteq C^i$ and $(x,d] \subseteq X \setminus C^i$. Evidently we must have $x \in H$. It follows first that $D^i \subseteq \langle H \cup c \rangle$ and then that $C^i \subseteq \langle H \cup c \rangle$. Hence $\langle H \cup c \rangle = X$ and H is a hyperplane. Since C^i and D^i are convex and partition $X \setminus H$, they are the open half-spaces associated with H. Hence, by Proposition V.28, $\overline{C} = C^i \cup H$ and $\overline{D} = D^i \cup H$. □

Let C, D be a convex partition of X. If X is finite-dimensional then, by Proposition V.16, $C^i \neq \emptyset$ and $D^i \neq \emptyset$. However, when X is infinite-dimensional it is possible that $C^i = D^i = \emptyset$, even though $\langle C \rangle = \langle D \rangle = X$. For example, take X to be the set of all sequences (x_n) of real numbers with at most finitely many non-zero terms. We can give X the structure of a real vector space by putting $(x_n) + (y_n) = (x_n + y_n)$ and $\lambda(x_n) = (\lambda x_n)$, and then define convexity in the usual way. Let C, C' be the subsets consisting of all sequences (x_n) whose last non-zero term is positive, negative, and put $D = C' \cup \{0\}$. Then C, D is a convex partition of X. However, for each point $x \in X$ there exist points $y,z \in X$, with $x \in (y,z)$, such that $(x,y] \subseteq C$ and $(x,z] \subseteq C'$. In fact, if $x = (x_n)$, where $x_n = 0$ for $n > m$, we can take $y = (y_n)$ and $z = (z_n)$, where $y_n = z_n = x_n$ for $n \neq m+1$ and $y_{m+1} = 1$, $z_{m+1} = -1$.

Under our present hypotheses there is a simple characterization of affine sets:

PROPOSITION 6 *A set $C \subseteq X$ is affine if and only if C is convex and $C^i = \overline{C}$.*

Proof Suppose first that C is affine. Then C is certainly convex and, since X is unending, $C^i = C$. It follows at once that also $\overline{C} = C$.

Suppose next that C is convex and $C^i = \overline{C}$. If C is not affine, there exist distinct points $c,c' \in C$ and $x \in X \setminus C$ such that $c' \in (c,x)$. By (S), there exists $y \in (c,x]$ such that $(c,y) \subseteq C$, $(y,x) \subseteq X \setminus C$. Since $C^i = C = \overline{C}$, it follows that $y \in C$, $y \neq x$ and (y,x) contains a point $c'' \in C$, which is a contradiction. □

Under our present hypotheses there is also a simple characterization of convex sets whose 'boundary' is convex:

PROPOSITION 7 *Let C be a convex set which is not affine. If $C^i \neq \emptyset$ and if $H := \overline{C} \setminus C^i$ is convex, then H is a hyperplane of $\langle C\rangle$, C^i is an associated open half-space and $\overline{C}$ is the corresponding closed half-space.*

Proof By Proposition 6, $H \neq \emptyset$. Since $\overline{C^i} = C^i$, it follows that $\overline{C}$ is not affine and thus $\overline{C} \subset \langle C\rangle$. Let $c \in C^i$ and $a \in \langle C\rangle \setminus \overline{C}$. Then, by Proposition V.9, there exists a point $c' \in (c,a) \cap C^i$. Hence, by (S), there exists a point $h \in (c,a]$ such that $(c,h) \subseteq C^i$ and $(h,a] \subseteq X \setminus C^i$. Then $h \in H$, since $h \in \overline{C}$, $h \neq a$ and hence $h \notin C^i$.

We now show that $D = \langle C\rangle \setminus C^i$ is convex. Assume on the contrary that there exists $c \in C^i$ such that $c \in (d,d')$ for some distinct $d,d' \in D$. If $d \in \overline{C}$ choose a so that $d \in (c,a)$, and if $d \notin \overline{C}$ put $a = d$. Similarly if $d' \in \overline{C}$ choose a' so that $d' \in (c,a')$, and if $d' \notin \overline{C}$ put $a' = d'$. Then $a,a' \notin \overline{C}$. By the argument of the preceding paragraph, the segments (c,a) and (c,a') each contain a point of H. Since H is convex, it follows that $c \in H$, which is a contradiction.

Thus C^i,D is a convex partition of $\langle C\rangle$. By Proposition 5 it now need only be shown that $H = D \setminus D^i$. As in the proof of Proposition 5, $\overline{C^i} = \langle C\rangle \setminus D^i$. But $\overline{C} = \overline{C^i}$, since $C^i \neq \emptyset$, and $H = \overline{C} \cap D$. Hence $H = D \setminus D^i$. □

Let H be a hyperplane of X and let S,T be nonempty subsets of X. The sets S and T are said to be *separated* by the hyperplane H if S is contained in one of the closed half-spaces associated with H and T is contained in the other. They are said to be *properly separated* by H if, in addition, $S \cup T \not\subseteq H$.

The case in which the sets S and T are convex is of particular interest. The following *first separation theorem* deals with this case:

PROPOSITION 8 *Let A and B be convex sets such that $A^i \neq \emptyset$, $\langle A \rangle = X$ and $B \neq \emptyset$. Then A and B can be separated by a hyperplane if and only if $A^i \cap B = \emptyset$, and A^i is then contained in one of the open half-spaces associated with this hyperplane.*

Proof Suppose first that A and B can be separated by a hyperplane H. Then A is not contained in H, since $\langle A \rangle = X$, and hence $A^i \cap H = \emptyset$. Thus A^i is contained in one of the open half-spaces associated with H and $A^i \cap B = \emptyset$.

Suppose next that $A^i \cap B = \emptyset$. Then, by Proposition II.22, there exists a convex partition C,D of X with $A^i \subseteq C, B \subseteq D$. In fact $A^i \subseteq C^i$, by Proposition V.10, since $\langle A^i \rangle = X = \langle C \rangle$. If we put $H = X \setminus (C^i \cup D^i)$ then, by Proposition 5, H is a hyperplane and C^i, D^i are the two open half-spaces associated with H. It follows at once that A and B are separated by H. □

In Proposition 8 the hypothesis $\langle A \rangle = X$ is rather strong and the convex sets A and B do not appear on an equal footing. These drawbacks are removed in the following deeper-lying *second separation theorem*:

PROPOSITION 9 *Let A and B be convex sets such that $A^i \neq \emptyset$ and $B^i \neq \emptyset$. Then A and B can be properly separated by a hyperplane of X if and only if $A^i \cap B^i = \emptyset$.*

Proof Suppose first that A and B are properly separated by a hyperplane H, and assume that there exists a point $x \in A^i \cap B^i$. Then $x \in H$ and there exists also a point $y \in (A \cup B) \setminus H$. Without loss of generality assume $y \in A \setminus H$. Since $x \in A^i$, there exists a point $z \in A$ such that $x \in (y,z)$. Then y and z lie in different open half-spaces associated with H, which is a contradiction.

Suppose next that $A^i \cap B^i = \emptyset$. We show first that if a hyperplane H properly separates A^i and B^i, then it also properly separates A and B. Indeed A^i and B^i are not both contained in H. If A^i, say, is not contained in H then, since $A^{ii} = A^i$, A^i is contained in an open half-space H_+ associated with H and B^i is contained in the disjoint closed half-space $H_- \cup H$. Moreover, since $B^{ii} = B^i$, either $B^i \subseteq H_-$ or $B^i \subseteq H$. It follows that $A \subseteq H_+ \cup H$ and $B \subseteq H_- \cup H$. Thus, by replacing A by A^i and B by B^i, we may now assume without loss of generality that $A = A^i$ and $B = B^i$.

Suppose $X' := \langle A \cup B \rangle \neq X$ and there exists a hyperplane H' of X' which properly separates A and B. Let $x' \in X' \backslash H'$. By Hausdorff's maximality theorem there exists an affine set H which contains H' but not x', and such that any affine set properly containing H does contain x'. Thus if $x \in X \backslash H$, then $x' \in \langle x \cup H \rangle$ and so, since $x' \notin H$, $x \in \langle x' \cup H \rangle$. Hence $\langle x' \cup H \rangle = X$ and H is a hyperplane of X. Evidently $H \cap X' = H'$. If H_+ and H_- are the open half-spaces of X associated with its hyperplane H, then $H_+ \cap X'$ and $H_- \cap X'$ are the open half-spaces of X' associated with its hyperplane H', since they are convex and partition $X' \backslash H'$. It follows that A and B are also properly separated by the hyperplane H of X. Thus we now assume, without loss of generality, that $X = \langle A \cup B \rangle$.

Put $C = A/B$. Then C is convex, by Proposition V.32, and $\langle C \rangle = \langle A \cup B \rangle = X$. Moreover $B \cap C = \emptyset$, since $A \cap B = \emptyset$, and $C^i = C$, by Proposition V.39. Hence, by Proposition 8, B and C can be separated by a hyperplane H of X and C is contained in an open half-space H_+ associated with H. Moreover, since $B^i = B$, B is contained in the other open half-space H_- or in H. In either case $A \cap H_- = \emptyset$, and $A \subseteq H_+$ if $B \subseteq H$. Thus H properly separates A and B. □

COROLLARY 10 *Two nonempty finite-dimensional convex sets A,B can be properly separated by a hyperplane if and only if* $A^i \cap B^i = \emptyset$. □

An important consequence of Proposition 9 is the following *extension theorem*, which reduces to the geometric form of the Hahn–Banach theorem in the case where X is a real vector space.

PROPOSITION 11 *If A is a convex set such that $A^i \neq \emptyset$ and if B is a nonempty affine set such that $A^i \cap B = \emptyset$, then there exists a hyperplane H such that $B \subseteq H$ and $A^i \cap H = \emptyset$.*

Proof Since B is affine, $B^i = B$. Hence, by Proposition 9, there exists a hyperplane H which properly separates A and B. If $B \subseteq H$, then $A^i \cap H = \emptyset$ and there is nothing more to do. Otherwise, since B is affine, B is contained in an open half-space H_- of X associated with the hyperplane H and A is contained in the complementary closed half-space $H_+ \cup H$.

Put $A' = H_+ \cup H$, and let C denote the set of all points c such that $c \in (a;b)$ for some $a \in A'$ and $b \in B$. Then, since B is affine, $B \cap C = \emptyset$ and $A' \subseteq C$. Moreover, C is convex. For suppose $c \in (c_1,c_2)$, where c_1 and c_2 are distinct elements of C. Then $c_i \in (a_i,b_i)$ for some $a_i \in A'$ and $b_i \in B$ ($i = 1,2$). Hence $c \in [a_1,a_2,b_1,b_2]$ and $c \in [a,b]$ for some $a \in [a_1,a_2]$ and $b \in [b_1,b_2]$. If $c = b$, then $c,c_1,c_2 \in \langle a_1,b_1,b_2\rangle$ and a_1,a_2,c_1,c_2 all lie in the same open half-plane associated with the line $\langle b_1,b_2\rangle$, which contradicts $c \in (c_1,c_2)$. Hence $c \neq b$ and $c \in C$. Furthermore $C^i \neq \emptyset$, since $H_+ \subseteq C^i$.

Consequently, by Proposition 8, there exists a hyperplane H' which separates B and C. Moreover $C^i \subseteq H_+'$, where H_+' is one of the open half-spaces associated with H', and $C \subseteq H_+' \cup H'$. It follows from the definition of C that $B \cap H_-' = \emptyset$, where H_-' is the other open half-space associated with H', and so $B \subseteq H'$.

Hence $H \neq H'$. Furthermore $H \cap H' = \emptyset$. For assume $a' \in H \cap H'$ and choose any $a \in H$ such that $a \neq a'$. Then $\langle a,a'\rangle \subseteq H$. Since $H \subseteq C$, it follows that $\langle a,a'\rangle \subseteq H'$. Thus $H \subseteq H'$, which is a contradiction.

It follows that $H \subseteq H_+'$, and hence also $A' = H_+ \cup H \subseteq H_+'$. □

PROPOSITION 12 *Let C be a convex set with $C^i \neq \emptyset$ and let $x \in \overline{C} \setminus C^i$. Then there exists at least one hyperplane H_x of X, with $x \in H_x$, such that C^i is contained in one of the open half-spaces associated with H_x and $\overline{C}$ is contained in the corresponding closed half-space.*

Moreover, $\overline{C}$ is the intersection of all such closed half-spaces and C^i the intersection of all such open half-spaces, for varying $x \in \overline{C} \setminus C^i$.

Proof By Proposition 8, x and C^i can be separated by a hyperplane H_x' of $\langle C\rangle$ so that C^i is contained in one of the open half-spaces associated with H_x'. Moreover, since $x \in \overline{C}$, we must have $x \in H_x'$. By Proposition 11 there exists a hyperplane H_x of X containing H_x' such that C^i is contained in one of the open half-spaces associated with H_x. Then $\overline{C}$ is contained in the corresponding closed half-space.

If D is the intersection of *all* such closed half-spaces, for varying $x \in \overline{C} \setminus C^i$, then obviously $\overline{C} \subseteq D$. Assume there exists a point $y \in D \setminus \overline{C}$ and choose $z \in C^i$. Then the segment $[y,z]$ contains a point x such that $[z,x) \subseteq C^i$ and $(x,y] \cap C^i = \emptyset$. Evidently $x \in \overline{C} \setminus C^i$. Hence there exists a hyperplane containing x such that C^i is contained in one of the associated open half-spaces. Since y is in the other associated open half-space, this is a contradiction. We conclude that $\overline{C} = D$. The corresponding characterization of C^i follows immediately. □

A hyperplane H_x with the properties described in Proposition 12 will be said to be a *supporting hyperplane* of C at x, and the associated closed half-space which contains $\overline{C}$ will be said to be a *supporting half-space* of C at x.

Evidently a hyperplane H_x of X which contains a point $x \in \overline{C}$ is a supporting hyperplane of C at x if and only if $\overline{C} \cap H_x$ is a proper face of $\overline{C}$. The intersection of all supporting half-spaces of C, at a fixed point $x \in \overline{C} \setminus C^i$, is a convex cone with vertex x, which will be called the *supporting cone* of C at x.

It follows at once from the definitions that if C is a convex cone with $C^i \neq \emptyset$, then any supporting hyperplane of C contains the affine set of vertices of C.

Finally we consider separation properties for any finite number of convex sets.

PROPOSITION 13 *Let $C_1,\ldots,C_n$ $(n > 1)$ be convex proper subsets of X such that $C_k^i \neq \emptyset$ $(1 \le k \le n)$ and*

$$C_1^i \cap \ldots \cap C_n^i = \emptyset.$$

Then there exist hyperplanes H_k with associated open half-spaces H_k^+, H_k^- such that $C_k \subseteq \overline{H_k^+}$ $(1 \le k \le n)$, $C_k \not\subseteq H_k$ for at least one k, and

$$H_1^+ \cap \ldots \cap H_n^+ = \emptyset.$$

Proof By Proposition 9 the result is true for $n = 2$, with $H_1 = H_2$ and $H_2^+ = H_1^-$. We assume that $n > 2$ and the result holds for all smaller values of n.

Suppose first that $C_2^i \cap \ldots \cap C_n^i = \emptyset$. Then, by the induction hypothesis, there exist hyperplanes H_k with associated open half-spaces H_k^+, H_k^- such that $C_k \subseteq \overline{H_k^+}$ $(2 \le k \le n)$, $C_k \not\subseteq H_k$ for at least one $k \ge 2$, and

$$H_2^+ \cap \ldots \cap H_n^+ = \emptyset.$$

Choose $x_1 \in X \setminus C_1$. By Proposition 11 there exists a hyperplane H_1 with associated open half-spaces H_1^+, H_1^- such that $x_1 \in H_1$ and $C_1^i \subseteq H_1^+$. Since $C_1 \subseteq \overline{H_1^+}$, the result certainly holds in this case. Thus we may now suppose $\bigcap_{k=1,k\neq j}^n C_k^i \neq \emptyset$ for each $j \in \{1,\ldots,n\}$.

If we put

$$B_1 = C_2 \cap \ldots \cap C_n$$

then, by Proposition V.12,

$$B_1^i = C_2^i \cap \ldots \cap C_n^i \neq \emptyset.$$

Hence, by Proposition 9, there exists a hyperplane H_1 which properly separates B_1 and C_1. Thus we may name the open half-spaces H_1^+, H_1^- associated with the hyperplane H_1 so that $C_1 \subseteq \overline{H_1^+}$ and $B_1 \subseteq \overline{H_1^-}$. Moreover, either $C_1^i \subseteq H_1^+$, or $C_1 \subseteq H_1$ and $B_1^i \subseteq H_1^-$.

Thus the hypotheses of the proposition remain satisfied if C_1 is replaced by $D_1 = \overline{H_1^+}$. By the induction hypothesis, in the same way as before, we may suppose that

$$D_1^i \cap \bigcap_{k=2, k\neq j}^{n} C_k^i \neq \emptyset \text{ for each } j \in \{2,\ldots,n\}.$$

If we put

$$B_2 = D_1 \cap C_3 \cap \ldots \cap C_n,$$

then $B_2^i \neq \emptyset$ and $B_2^i \cap C_2^i = D_1^i \cap B_1^i = \emptyset$. Hence there exists a hyperplane H_2 which properly separates B_2 and C_2. Thus the open half-spaces H_2^+, H_2^- associated with the hyperplane H_2 may be named so that $C_2 \subseteq \overline{H_2^+}$ and $B_2 \subseteq \overline{H_2^-}$. Moreover, either $C_2^i \subseteq H_2^+$ or $C_2 \subseteq H_2$ and $B_2^i \subseteq H_2^-$.

Thus the hypotheses of the proposition remain satisfied if also C_2 is replaced by $D_2 = \overline{H_2^+}$. Proceeding in this way we define inductively for $k = 1,\ldots,n$ hyperplanes H_k with associated open half-spaces H_k^+, H_k^- such that $C_k \subseteq D_k$ and $B_k \subseteq \overline{H_k^-}$, where $D_k = \overline{H_k^+}$ $(1 \leq k \leq n)$,

$$B_k = D_1 \cap \ldots \cap D_{k-1} \cap C_{k+1} \cap \ldots \cap C_n \quad (1 \leq k < n)$$

and

$$B_n = D_1 \cap \ldots \cap D_{n-1}.$$

Moreover, either $C_k^i \subseteq D_k^i = H_k^+$ or $C_k \subseteq H_k$ and $B_k^i \subseteq H_k^-$. Furthermore, although

$$B_k^i \cap C_k^i = D_1^i \cap \ldots \cap D_{k-1}^i \cap C_k^i \cap \ldots \cap C_n^i \quad (1 \leq k < n)$$

is empty, all intersections obtained by omitting one term on the right are nonempty. Since

$$H_1^+ \cap \ldots \cap H_n^+ \subseteq B_n \cap H_n^+ = \emptyset,$$

the proof is complete unless we have $C_k \subseteq H_k$ for every $k \in \{1,\ldots,n\}$.

In this case all intersections obtained by omitting one term from $H_1 \cap \ldots \cap H_n$ are nonempty and, for $1 < k \leq n$, all intersections obtained by omitting one term from

$$H_1^+ \cap \ldots \cap H_{k-1}^+ \cap H_k \cap \ldots \cap H_n$$

are nonempty. We are going to show that $H_1 \cap \ldots \cap H_n = \emptyset$.

Assume on the contrary that there exists a point $z \in H_1 \cap \ldots \cap H_n$ and let $y \in B_1^i$. Then $y \in H_1^- \cap H_2 \cap \ldots \cap H_n$. If we choose x so that $z \in (x,y)$, then $x \in H_1^+ \cap H_2 \cap \ldots \cap H_n$. Let $u \in H_2^+ \cap \ldots \cap H_{n-1}^+ \cap H_n$. Then $(u,x) \subseteq H_2^+ \cap \ldots \cap H_{n-1}^+ \cap H_n$. But, since $x \in H_1^+$, we can choose $v \in (u,x)$ so that $v \in H_1^+$. Then $v \in H_1^+ \cap \ldots \cap H_{n-1}^+ \cap H_n$, which contradicts $B_n^i \subseteq H_n^-$.

It now follows from Proposition 11 that there exists a hyperplane L_n with associated open half-spaces L_n^+, L_n^- such that $H_1 \cap \ldots \cap H_{n-1} \subseteq L_n$ and $H_n \subseteq L_n^+$.

If $z \in H_1^+ \cap H_2 \cap \ldots \cap H_{n-1}$ and $y \in B_1^i$, then $y \in H_1^- \cap H_2 \cap \ldots \cap H_n$ and there exists a point $x \in (y,z)$ such that $x \in H_1 \cap \ldots \cap H_{n-1}$. Since $x \in L_n$ and $y \in L_n^+$, it follows that $z \in L_n^-$. Thus $H_1^+ \cap H_2 \cap \ldots \cap H_{n-1} \subseteq L_n^-$.

Suppose that, for some k with $1 < k \leq n-1$, we have

$$H_1^+ \cap \ldots \cap H_{k-1}^+ \cap H_k \cap \ldots \cap H_{n-1} \subseteq L_n^-.$$

We have just shown that this is true for $k = 2$ and we wish to show that if it holds as written, then it also holds when k is replaced by $k+1$. Let

$$x \in H_1^+ \cap \ldots \cap H_k^+ \cap H_{k+1} \cap \ldots \cap H_{n-1}$$

and assume that $x \in L_n$. Let $y \in B_k^i$, so that

$$y \in H_1^+ \cap \ldots \cap H_{k-1}^+ \cap H_k^- \cap H_{k+1} \cap \ldots \cap H_n,$$

and let

$$z \in H_1^+ \cap \ldots \cap H_{k-1}^+ \cap H_k \cap \ldots \cap H_{n-1}.$$

Since $H_n \subseteq L_n^+$ and $z \in L_n^-$, by supposition, there exists a point $w \in (y,z)$ such that $w \in L_n$. Then

$$w \in H_1^+ \cap \ldots \cap H_{k-1}^+ \cap H_k^- \cap H_{k+1} \cap \ldots \cap H_{n-1} \cap L_n.$$

Hence there exists $z' \in (x,w)$ such that $z' \in H_k$. Then

$$z' \in H_1^+ \cap \ldots \cap H_{k-1}^+ \cap H_k \cap \ldots \cap H_{n-1} \cap L_n,$$

which is contrary to our supposition. We conclude that

$$H_1^+ \cap \ldots \cap H_k^+ \cap H_{k+1} \cap \ldots \cap H_{n-1} \cap L_n = \emptyset.$$

Since $(x,z) \subseteq H_1^+ \cap \ldots \cap H_k^+ \cap H_{k+1} \cap \ldots \cap H_{n-1}$, we must actually have

$$H_1^+ \cap \ldots \cap H_k^+ \cap H_{k+1} \cap \ldots \cap H_{n-1} \subseteq L_n^-.$$

Hence it follows by induction that

$$H_1^+ \cap \ldots \cap H_{n-1}^+ \subseteq L_n^-,$$

and thus $H_1^+ \cap \ldots \cap H_{n-1}^+ \cap L_n^+ = \emptyset$. □

3 FUNDAMENTAL THEOREM OF ORDERED GEOMETRY

An ordered division ring D is a one-dimensional dense unending linear geometry. It will now be shown that D is complete (if and) only if D is the field $\mathbb{R}$ of real numbers.

PROPOSITION 14 *Let D be an ordered division ring. If, for any convex set $C \subseteq [0,1]$ with $0 \in C$ and $1 \notin C$, there exists $c \in [0,1]$ such that $[0,c) \subseteq C$ and $(c,1] \subseteq D \setminus C$, then there is a bijective map of D onto the real field $\mathbb{R}$ which preserves sums, products and order.*

Proof The interval $[0,1]$ may be linearly mapped onto the interval $[a,b]$, for any $a,b \in D$ with $a \neq b$. Consequently the property in the statement of the proposition continues to hold if 0 is replaced by a and 1 by b.

In any ordered division ring D, $n1 = 1 + \ldots + 1$ (n times) is positive for any positive integer n. It follows that the set of all elements $m1\cdot(n1)^{-1}$, where m is any integer and n is any positive integer, is a field isomorphic to the field $\mathbb{Q}$ of rational numbers. By identifying $m1\cdot(n1)^{-1}$ with mn^{-1}, we may regard $\mathbb{Q}$ as embedded in D.

We show first that for each $a \in D$ there exists a positive integer n such that $n > a$. Assume on the contrary that $n \leq a$ for every positive integer n. Then $a > 0$ and the set C of all $b \in [0,a]$ such that $n \leq b$ for every positive integer n is convex and contains a, but not 0. Hence there exists $c \in [0,a]$ such that $(c,a] \subseteq C$ and $[0,c) \subseteq D \setminus C$. Moreover $1 < c$, since $2 \notin C$. Thus if we choose $d \in D$ so that $1 + d = c$, then $0 < d < c$. Since $d \notin C$, we have $m > d$ for some positive integer m. Hence $1 + m > 1 + d = c$, which is a contradiction.

Consequently, for any $a,b \in D$ with $a > 0$, there is a positive integer n such that $n > ba^{-1}$, and then $na > b$.

We show next that, for any $a,b \in D$ with $b > a$, there exists $x \in \mathbb{Q}$ such that $b > x > a$. Let n be a positive integer such that $n(b-a) > 1$ and let m be the least integer such that $m > na$. Then $nb > na + 1 \geq (m-1) + 1 = m > na$ and hence $b > mn^{-1} > a$.

It will now be shown that D is a field. It is sufficient to show that $ab = ba$ for all $a > 0$, $b > 0$. Assume on the contrary that there exist $a > 0$, $b > 0$ such that $ab > ba$. Then $ab > x > ba$ for some $x \in \mathbb{Q}$. Thus $b^{-1}x > a$ and $a > xb^{-1}$, hence $b^{-1}x > xb^{-1}$. Then $b^{-1}x > y > xb^{-1}$ for some $y \in \mathbb{Q}$. Thus $b^{-1} > yx^{-1}$ and $x^{-1}y > b^{-1}$. Since $yx^{-1} = x^{-1}y$, this is a contradiction.

For any $a \in D$, the set of all $x \in \mathbb{Q}$ with $x < a$ is a 'cut' in $\mathbb{Q}$. Moreover, the hypothesis of the proposition ensures that all cuts in $\mathbb{Q}$ may be obtained in this way. It is readily verified that the mapping which makes correspond to a the real number defined by this cut in Dedekind's construction of the reals from the rationals has all the required properties. □

COROLLARY 15 *In the statements of Theorems VII.3,4 and 5 the ordered division ring D may be replaced by the real field $\mathbb{R}$ if the linear geometry X is also assumed to be complete.* □

Every convex subset C of a real vector space such that $C^i = C$ is a complete unending linear geometry. The following theorem essentially characterizes such sets. Since in a sense it is the culmination of our work, we designate it the *fundamental theorem of ordered geometry.*

THEOREM 16 *If X is a complete unending linear geometry with* $\dim X > 2$, *or with* $\dim X = 2$ *and the Desargues property, then X is isomorphic to a convex subset C of a real vector space such that $C^i = C$.*

Proof By Theorem VII.4 and Corollary 15, there is an isomorphism f of X with a set C in the projective completion $\overline{V}$ of a real vector space V. Moreover $C^i = C$, since $X^i = X$.

Let a_0,b_0 be distinct points of X. By Proposition V.18, there exists a hyperplane H of X such that a_0,b_0 lie in different open half-spaces H_+,H_- of X associated with the hyperplane H. The images Y,Y_+,Y_- of H,H_+,H_- under the isomorphism f are disjoint nonempty convex sets with union C. Moreover $Y_+^i = Y_+$ and $Y_-^i = Y_-$, since $H_+^i = H_+$ and $H_-^i = H_-$.

The theorem will be proved if we show that there is a 'hyperplane' of $\overline{V}$ disjoint from C. Since it is sufficient to show that there is a 'hyperplane' disjoint from C in the projective subspace of $\overline{V}$ generated by C, we may assume that C itself generates $\overline{V}$. Then there is a unique 'hyperplane' K of $\overline{V}$ containing Y and Y_+, Y_- are contained in the affine space $A = \overline{V} \setminus K$. Since A is the affine hull of Y_+ and of Y_-, it follows from Proposition 8 that there is a hyperplane L of A such that Y_+ and Y_- are contained in different open half-spaces of A associated with the hyperplane L. The hyperplane L of the affine space A extends to a 'hyperplane' $\overline{L}$ of $\overline{V}$. Evidently Y_+ and Y_- are disjoint from $\overline{L}$, since $\overline{L}$ is obtained from L by adjoining points of K. It only remains to show that Y is also disjoint from $\overline{L}$. Suppose $c' \in Y$, so that $c' = f(c)$ for some $c \in H$. Then $c \in (a,b)$ for some $a \in H_+, b \in H_-$. If $a' = f(a)$, $b' = f(b)$, then the line $\langle a',b' \rangle$ of the affine space A contains a point of L. Since $c' \notin L$ and $a',b' \notin \overline{L}$, it follows that $c' \notin \overline{L}$. □

The axioms actually required for Theorem 16 are **(C)**,**(L1)**–**(L4)**,**(U)** and **(S)**, since **(P)** and **(D)** are then implied by dim $X > 1$. We now consider the necessity of the hypotheses in Theorem 16. Example VI.7 shows that Theorem 16 no longer holds if in its statement we replace 'unending' by 'dense', even if we omit the requirement that $C^i = C$. Examples VI.4 and VI.5 show that Theorem 16 no longer holds when dim $X = 2$ if we omit the requirement that X have the Desargues property. Furthermore Example 1 of this chapter shows the necessity of excluding the case dim $X = 1$. Finally we are going to show that in the statement of Theorem 16 we cannot omit 'complete' and replace 'real vector space' by 'vector space over an ordered division ring'.

We first prove

LEMMA 17 *The real field* $\mathbb{R}$ *has no proper subfield of which it is a finite extension.*

Proof It is sufficient to show that if K is a subfield of the complex field $\mathbb{C}$ such that $[\mathbb{C} : K] = n$, where $1 < n < \infty$, then $n = 2$. Evidently $\mathbb{C}$ is a normal extension of K, since any polynomial in $K[x]$ is a product of linear factors in $\mathbb{C}[x]$ by the fundamental theorem of algebra. Since the Galois group G of $\mathbb{C}$ over K has order n, it has an element g of order p, where p is some prime divisor of n. By the fundamental theorem of Galois theory, the elements of $\mathbb{C}$ fixed by the automorphism g form a field F such that $K \subseteq F \subseteq \mathbb{C}$ and $[\mathbb{C} : F] = p$. Since

$$x^p - 1 = (x-1)(1 + x + ... + x^{p-1}),$$

and since an irreducible polynomial in $F[x]$ has degree dividing $[\mathbb{C} : F] = p$, the irreducible factors of $x^p - 1$ in $F[x]$ are all linear. Thus F contains a primitive p-th root of unity ζ.

We are going to show that $\mathbb{C} = F(\alpha)$, where $\alpha^p \in F$. Let $\gamma \in \mathbb{C} \setminus F$, so that $\mathbb{C} = F(\gamma)$. If

$$\alpha = \gamma + \zeta g(\gamma) + ... + \zeta^{p-1} g^{p-1}(\gamma),$$

then

$$g(\alpha) = g(\gamma) + \zeta g^2(\gamma) + ... + \zeta^{p-1} g^p(\gamma) = \zeta^{-1}\alpha.$$

Thus $\alpha \notin F$, since $g(\alpha) \neq \alpha$. But $g(\alpha^p) = \zeta^{-p}\alpha^p = \alpha^p$. Hence $\alpha^p \in F$ and $\mathbb{C} = F(\alpha)$.

Choose $\beta \in \mathbb{C}$ so that $\beta^p = \alpha$. If $b = \prod_{j=1}^{p} g^{j-1}(\beta)$, then $g(b) = b$ and hence $b \in F$. Moreover $b^p = \prod_{j=1}^{p} g^{j-1}(\alpha)$ and $g^{j-1}(\alpha) = \zeta^{-j+1}\alpha$ $(j = 1,...,p)$. If $p \neq 2$, then $1 + 2 + ... + (p-1) = p(p-1)/2$ is divisible by p and hence $b^p = \alpha^p$. Since $\alpha \notin F$, $b \in F$ and F contains all p-th roots of unity, this is a contradiction. Hence $p = 2$, $\zeta = -1$ and $b^2 = -\alpha^2$. Since $(\mathrm{i}b)^2 = \alpha^2$ and $\alpha \notin F$, it follows that $\mathrm{i} \notin F$. *A fortiori*, $\mathrm{i} \notin K$. If $K(\mathrm{i}) \neq \mathbb{C}$, the same argument can be applied with $K(\mathrm{i})$ in place of K to obtain the contradiction $\mathrm{i} \notin K(\mathrm{i})$. Hence $K(\mathrm{i}) = \mathbb{C}$ and $n = 2$. □

PROPOSITION 18 *For any ordered division ring D which is not isomorphic to the field $\mathbb{R}$ of real numbers and any positive integer n, the projective space $P^n(D)$ can be given the structure of a dense, unending linear geometry.*

Proof The proof will be based on simple facts about filters. Let X be a nonempty set. A nonempty collection $\mathcal{F}$ of subsets of X is said to be a *filter* on X when the following properties hold:

(F1) *if $A \in \mathcal{F}$ and $A \subseteq B \subseteq X$, then $B \in \mathcal{F}$;*
(F2) *if $A \in \mathcal{F}$ and $B \in \mathcal{F}$, then $A \cap B \in \mathcal{F}$;*
(F3) $\emptyset \notin \mathcal{F}$.

Properties **(F2)** and **(F3)** imply that the intersection of finitely many sets in $\mathcal{F}$ is nonempty and, since $\mathcal{F}$ is nonempty, property **(F1)** implies that $X \in \mathcal{F}$.

If $\mathcal{F}$ and $\mathcal{F}'$ are filters on X, we write $\mathcal{F} \subseteq \mathcal{F}'$ if $A \in \mathcal{F}$ implies $A \in \mathcal{F}'$. Evidently this is a partial ordering of the set of all filters on X. An *ultrafilter* on X is a filter which is a maximal element in this partial ordering. It follows from Hausdorff's maximality theorem that, for any filter $\mathcal{F}$ on X, there exists an ultrafilter $\mathcal{U}$ on X such that $\mathcal{F} \subseteq \mathcal{U}$.

If $\mathcal{U}$ is an ultrafilter on X and A,B are subsets of X such that $A \cup B \in \mathcal{U}$, then either $A \in \mathcal{U}$ or $B \in \mathcal{U}$ (or both). For assume on the contrary that $A,B \notin \mathcal{U}$. Then the collection $\mathcal{F}$ of all sets $F \subseteq X$ such that $A \cup F \in \mathcal{U}$ is a filter on X. Since $\mathcal{U} \subseteq \mathcal{F}$ and $B \in \mathcal{F}$, this contradicts the hypothesis that $\mathcal{U}$ is an ultrafilter.

It follows that if $\mathcal{U}$ is an ultrafilter on X then, for any $C \subseteq X$, either $C \in \mathcal{U}$ or $X \backslash C \in \mathcal{U}$. We will show that, conversely, if $\mathcal{F}$ is a filter on X such that, for any $C \subseteq X$, either $C \in \mathcal{F}$ or $X \backslash C \in \mathcal{F}$, then $\mathcal{F}$ is an ultrafilter. Indeed otherwise there exists $E \subseteq X$ such that $E \notin \mathcal{F}$ and $E \cap F \neq \emptyset$ for every $F \in \mathcal{F}$. Since $E \notin \mathcal{F}$ implies $X \backslash E \in \mathcal{F}$ and $E \cap (X \backslash E) = \emptyset$, this is a contradiction.

Now let D be an ordered division ring, $\mathcal{U}$ an ultrafilter on a set X, and Γ the set of all maps $\phi: X \to D$. If $\phi,\phi' \in \Gamma$, we write $\phi \sim \phi'$ if the set $\{x \in X: \phi(x) = \phi'(x)\} \in \mathcal{U}$. It is easily verified that this is an equivalence relation on Γ. The set of equivalence classes will be denoted by $\tilde{\Gamma}(X,\mathcal{U})$, or simply $\tilde{\Gamma}$, since X and $\mathcal{U}$ will be fixed. If $\tilde{\phi}, \tilde{\psi}, \tilde{\omega} \in \tilde{\Gamma}$, we define $\tilde{\omega} = \tilde{\phi} + \tilde{\psi}$ if there exist $\phi \in \tilde{\phi}$, $\psi \in \tilde{\psi}$, $\omega \in \tilde{\omega}$ such that the set $\{x \in X: \omega(x) = \phi(x) + \psi(x)\} \in \mathcal{U}$. This definition does not depend on the choice of ϕ,ψ,ω within their equivalence classes. For if $\phi' \sim \phi$, $\psi' \sim \psi$, $\omega' \sim \omega$, then the set $\{x \in X: \omega'(x) = \phi'(x) + \psi'(x)\}$ contains the intersection of the sets $\{x \in X: \phi'(x) = \phi(x)\}$, $\{x \in X: \psi'(x) = \psi(x)\}$, $\{x \in X: \omega'(x) = \omega(x)\}$, $\{x \in X: \omega(x) = \phi(x) + \psi(x)\}$ and consequently, since $\mathcal{U}$ is a filter, is in $\mathcal{U}$.

If $\tilde{\phi}, \tilde{\psi}, \tilde{\omega} \in \tilde{\Gamma}$, we may similarly define $\tilde{\omega} = \tilde{\phi} \cdot \tilde{\psi}$ if there exist $\phi \in \tilde{\phi}$, $\psi \in \tilde{\psi}$, $\omega \in \tilde{\omega}$ such that the set $\{x \in X: \omega(x) = \phi(x) \cdot \psi(x)\} \in \mathcal{U}$. We may also define $\tilde{\phi} \leq \tilde{\psi}$ if there exist $\phi \in \tilde{\phi}$, $\psi \in \tilde{\psi}$ such that the set $\{x \in X: \phi(x) \leq \psi(x)\} \in \mathcal{U}$. Let $\tilde{0}, \tilde{1}$ be the equivalence classes in $\tilde{\Gamma}$ containing the constant maps $\phi(x) = 0$, $\phi(x) = 1$ respectively, for every $x \in X$. It is not difficult to verify that with these definitions $\tilde{\Gamma}$ is an ordered division ring. We give the argument only when it depends on the fact that $\mathcal{U}$ is actually an ultrafilter.

Suppose $\tilde{\phi} \in \tilde{\Gamma}$ and $\tilde{\phi} \neq \tilde{0}$. We wish to show that there exists $\tilde{\psi} \in \tilde{\Gamma}$ such that $\tilde{\phi} \cdot \tilde{\psi} = \tilde{\psi} \cdot \tilde{\phi} = \tilde{1}$. If $\phi \in \tilde{\phi}$, then the set $\{x \in X: \phi(x) = 0\} \notin \mathcal{U}$. Since $\mathcal{U}$ is an ultrafilter, it follows that the set $\{x \in X: \phi(x) \neq 0\} \in \mathcal{U}$. Define $\psi: X \to D$ by $\psi(x) = \phi(x)^{-1}$ if $\phi(x) \neq 0$, $= 1$ otherwise. If $\psi \in \tilde{\psi}$, then $\tilde{\phi} \cdot \tilde{\psi} = \tilde{\psi} \cdot \tilde{\phi} = \tilde{1}$.

Suppose $\tilde{\phi}, \tilde{\psi} \in \tilde{\Gamma}$ and $\tilde{\phi} \nleq \tilde{\psi}$. We wish to show that $\tilde{\psi} \leq \tilde{\phi}$. If $\phi \in \tilde{\phi}$, $\psi \in \tilde{\psi}$, then the set $\{x \in X: \phi(x) \leq \psi(x)\} \notin \mathcal{U}$. Since $\mathcal{U}$ is an ultrafilter, it follows that the set $\{x \in X: \psi(x) < \phi(x)\} \in \mathcal{U}$, and hence also the set $\{x \in X: \psi(x) \leq \phi(x)\} \in \mathcal{U}$. Thus $\tilde{\psi} \leq \tilde{\phi}$.

For any $a \in D$, define $\phi_a: X \to D$ by $\phi_a(x) = a$ for every $x \in X$. If $\phi_a \in \{\tilde{\phi}_a\}$, then the map $a \to \tilde{\phi}_a$ is an injection of D into $\tilde{\Gamma}$ which preserves addition, multiplication and order. Consequently we may regard D as embedded in $\tilde{\Gamma}$.

Suppose now that the ordered division ring D has a *gap*; i.e., there exist nonempty subsets A,B of D such that $a \leq b$ for every $a \in A$, $b \in B$, but there is no $c \in D$ such that $a \leq c \leq b$ for every $a \in A$, $b \in B$. This implies that $A \cap B = \emptyset$. We wish to construct an ordered division ring $D' \supset D$ in which the gap has been removed, i.e. there exists $c' \in D'$ such that $a \leq c' \leq b$ for every $a \in A$, $b \in B$.

For any $a \in A$, let $F_a = \{x \in A: a < x\}$ and let $\mathcal{F}$ denote the collection of all sets $F \subseteq A$ which contain a set F_a for some $a \in A$. Evidently $\mathcal{F}$ is a filter on A. Let $\mathcal{U}$ be an ultrafilter on A such that $\mathcal{F} \subseteq \mathcal{U}$, and now take $\tilde{\Gamma} = \tilde{\Gamma}(A,\mathcal{U})$ to be the ordered division ring constructed with $X = A$ and with this ultrafilter. Define the map $\gamma: A \to D$ by $\gamma(x) = x$ for every $x \in A$ and let $\tilde{\gamma}$ be the corresponding element of $\tilde{\Gamma}$. For each $a \in A$, the set $\{x \in A: a \leq \gamma(x)\} \in \mathcal{U}$, and for each $b \in B$, the set $\{x \in A: \gamma(x) \leq b\} = A \in \mathcal{U}$. Thus, regarding D as embedded in $\tilde{\Gamma}$, we have $a \leq \tilde{\gamma} \leq b$ for every $a \in A$, $b \in B$.

After these preparations we can proceed with the proof of Proposition 18. Assume first that D is a proper subfield of $\mathbb{R}$. Then, by Lemma 17, $\mathbb{R}$ is infinite-dimensional, considered as a vector space over D. Hence, by the discussion of Example VI.6, for any positive integer n the n-dimensional projective space over D can be given the structure of a dense, unending linear geometry.

Thus we now assume that no extension of D is isomorphic to the real field $\mathbb{R}$. Since D itself is not isomorphic to $\mathbb{R}$, it follows from Proposition 14 that the interval $[0,1]$ is the union of two nonempty subsets A,B such that $a \leq b$ for every $a \in A$, $b \in B$, but there is no $c \in D$ such that $a \leq c \leq b$ for every $a \in A$, $b \in B$. Hence, as we have seen, there exists an ordered division ring $D' \supset D$ with an element c' such that $a \leq c' \leq b$ for every $a \in A$, $b \in B$. Since no extension of D is isomorphic to $\mathbb{R}$, this argument can be repeated any finite number of times.

Consequently, for any positive integer n, there exists an ordered division ring $D^{(n)} \supset D$ which has dimension $> n$ as a vector space over D. The conclusion now follows as in the previous case. □

4 METRIC AND NORM

For convenience of reference we repeat here some well-known definitions concerning topological and metric spaces, and state some well-known results. The proofs may be found, for example, in Hewitt and Stromberg (1975).

A *topology* is said to be defined on a set X, and X is said to be a *topological space*, if a collection $\mathcal{T}$ of subsets of X is given such that

(T1) *the whole set X and the empty set $\varnothing$ are in $\mathcal{T}$,*
(T2) *the intersection of two sets in $\mathcal{T}$ is again a set in $\mathcal{T}$,*
(T3) *the union of any family of sets in $\mathcal{T}$ is again a set in $\mathcal{T}$.*

A set $A \subseteq X$ is said to be *open* if $A \in \mathcal{T}$ and is said to be *closed* if $X \backslash A \in \mathcal{T}$. A topological space is a *Hausdorff space* if any two distinct points are contained in disjoint open sets.

An *open cover* of a topological space X is a family $\{G_\alpha\}$ of open sets such that $X = \bigcup_\alpha G_\alpha$. A *subcover* is a subfamily which is also a cover. A topological space X is said to be *compact* if each open cover has a finite subcover.

Given a subset A of a topological space X, the collection of all sets $G \cap A$, where G is any open subset of X, defines a topology on A. If A is closed and X is compact, then A is also a compact topological space with the topology induced by that on X.

A *metric* is defined on a set X if with each ordered pair (x,y) of elements of X there is associated a real number $\mathrm{d}(x,y)$ with the properties

(D1) $\mathrm{d}(x,y) \geq 0$, *with equality if and only if $x = y$,*
(D2) $\mathrm{d}(x,y) = \mathrm{d}(y,x)$ *for all $x,y \in X$,*
(D3) $\mathrm{d}(x,y) \leq \mathrm{d}(x,z) + \mathrm{d}(z,y)$ *for all $x,y,z \in X$.*

The pair (X,d) is a *metric space*. A subset A of a metric space (X,d) is said to be *bounded* if $\{\mathrm{d}(x,y): x,y \in A\}$ is a bounded subset of $\mathbb{R}$.

A metric space (X,d) has a natural topology in which the open sets are the unions of sets of the form

$$N(x_0,\varepsilon) = \{x \in X: \mathrm{d}(x_0,x) < \varepsilon\},$$

where $x_0 \in X$ and $\varepsilon > 0$. This topology will always be understood when speaking of open or closed subsets of a metric space. For a metric space there is a simpler characterization of compactness: a metric space X is compact if and only if every sequence $\{x_n\}$ of elements of X has a convergent subsequence $\{x_{n_k}\}$, i.e. there exists $x \in X$ such that $\mathrm{d}(x_{n_k},x) \to 0$ as $k \to \infty$.

If a subset A of a metric space X is compact, with the topology induced by that on X, then A is closed and bounded. In general it is not true that, conversely, a closed and bounded set is compact, but it is true for $X = \mathbb{R}^d$.

A sequence $\{x_n\}$ in a metric space X is said to be a *Cauchy sequence* if for each real $\varepsilon > 0$ there is a corresponding integer $p > 0$ such that $\mathrm{d}(x_m,x_n) < \varepsilon$ for all $m,n \geq p$. A metric space is said to be *complete* if every Cauchy sequence is convergent. Any compact metric space is complete.

The two uses of the word 'complete' in this chapter correspond to the two ways, due to Dedekind and Cantor, of constructing the real numbers from the rationals. It is hoped that the meaning in each instance will be clear from the context.

Suppose now that V is a real vector space. We say that a *norm* is defined on V, and that V is a *normed vector space*, if with each $v \in V$ there is associated a real number $\|v\|$ with the properties

(N1) $\|v\| \geq 0$, *with equality if and only if* $v = 0$,
(N2) $\|v + w\| \leq \|v\| + \|w\|$ *for all* $v,w \in V$,
(N3) $\|\alpha v\| = |\alpha|\,\|v\|$ *for all* $v \in V$ *and all* $\alpha \in \mathbb{R}$.

It follows that V is a metric space with the metric $\mathrm{d}(v,w) = \|v - w\|$.

The following two propositions will be proved, since the proofs are less accessible.

PROPOSITION 19 *A finite-dimensional real vector space V can be normed. Moreover, whatever the norm, any bounded closed subset is compact.*

Proof Let d be the dimension of the vector space V and let $e_1,\ldots,e_d$ be a (vector space) basis for V. Then each $v \in V$ can be uniquely represented in the form

$$v = \alpha_1 e_1 + \ldots + \alpha_d e_d,$$

where $\alpha_j \in \mathbb{R}$ $(j = 1,\ldots,d)$. It may be immediately verified that

$$|v| = \max_{1\le j\le d} |\alpha_j|$$

is a norm on V. Moreover, the unit ball $U = \{v \in V: |v| \le 1\}$ is compact, since the interval $[-1,1]$ of $\mathbb{R}$ is compact. The unit sphere $S = \{v \in V: |v| = 1\}$ is also compact, since it is a closed subset of U.

To show that, with respect to an arbitrary norm $\| \ \|$, any bounded closed set is compact, it is sufficient to show that this norm is *equivalent* to the norm $| \ |$, i.e. there exist positive constants μ,ρ such that $\rho|v| \le \|v\| \le \mu|v|$ for every $v \in V$.

This is certainly true if $d = 1$, since then $\|v\| = |\alpha_1| \, \|e_1\| = \|e_1\| \, |v|$. We suppose $d > 1$ and use induction on d. Write $v = v' + v''$, where

$$v' = \alpha_1 e_1 + \ldots + \alpha_{d-1} e_{d-1}, \quad v'' = \alpha_d e_d.$$

By the induction hypothesis there exists a positive constant μ' such that, for every $v \in V$, $\|v'\| \le \mu'|v'|$. Then

$$\|v\| \le \|v'\| + \|v''\| \le \mu'|v'| + |\alpha_d| \, \|e_d\| \le \mu|v|,$$

where $\mu = \mu' + \|e_d\|$.

The real-valued function $\varphi(v) = \|v\|$ is continuous on S, since

$$|\varphi(v_1) - \varphi(v_2)| \le \|v_1 - v_2\| \le \mu|v_1 - v_2|.$$

Let $\rho = \inf\{\varphi(v): v \in S\}$ and let $\{v_n\}$ be a sequence of elements of S such that $\varphi(v_n) \to \rho$. Since S is compact, by restricting attention to a subsequence we may suppose that there exists $v^* \in S$ such that $|v_n - v^*| \to 0$. Then $\varphi(v^*) = \rho$, since φ is continuous, and $\rho > 0$, since $v^* \ne 0$. Since $\|v\| \ge \rho$ for $|v| = 1$, it follows from (N3) that $\rho|v| \le \|v\|$ for every $v \in V$. □

PROPOSITION 20 *If V is a normed real vector space, then a metric* d *can be defined on the corresponding projective space $P(V)$. Moreover, the metric space $(P(V),\mathrm{d})$ is complete if V is complete and compact if V is finite-dimensional.*

Proof The elements of $P(V)$ are the one-dimensional vector subspaces of V. If $x,y \in P(V)$ and if u,v are non-zero vectors in x,y respectively, we put

$$d(x,y) = \min\{\|(u/\|u\| - v/\|v\|)\|, \|(u/\|u\| + v/\|v\|)\|\}.$$

Evidently this definition does not depend on the choice of $u \in x$ and $v \in y$. Moreover $d(x,y) = d(y,x)$ and $d(x,y) \geq 0$, with equality if and only if $x = y$. Furthermore, the triangle inequality

$$d(x,y) \leq d(x,z) + d(z,y)$$

holds since, for any non-zero vectors u,v,w,

$$\|(u/\|u\| - v/\|v\|)\| \leq \|(u/\|u\| - w/\|w\|)\| + \|(w/\|w\| - v/\|v\|)\|,$$

$$\|(u/\|u\| - v/\|v\|)\| \leq \|(u/\|u\| + w/\|w\|)\| + \|(w/\|w\| + v/\|v\|)\|,$$

$$\|(u/\|u\| + v/\|v\|)\| \leq \|(u/\|u\| - w/\|w\|)\| + \|(w/\|w\| + v/\|v\|)\|,$$

$$\|(u/\|u\| + v/\|v\|)\| \leq \|(u/\|u\| + w/\|w\|)\| + \|(w/\|w\| - v/\|v\|)\|.$$

The projective space $P(V)$ is bounded with respect to this metric, since $d(x,y) \leq 2$ for all $x,y \in P(V)$.

Suppose now that V is complete and let $\{x_n\}$ be a Cauchy sequence in $P(V)$. Then there exists a positive integer p such that $d(x_n,x_p) < 1/2$ for every $n > p$. Choose a vector $u_n \in x_n$ with $\|u_n\| = 1$. By replacing u_n by $-u_n$, if necessary, we may assume that

$$d(x_n,x_p) = \|u_n - u_p\| \text{ for every } n > p.$$

Then, if $m,n > p$,

$$\|u_m - u_n\| \leq \|u_m - u_p\| + \|u_n - u_p\| < 1$$

and hence

$$\|u_m + u_n\| \geq 2\|u_m\| - \|u_m - u_n\| > 1.$$

Consequently $d(x_m,x_n) = \|u_m - u_n\|$ and $\{u_n\}$ is a Cauchy sequence in V. Since V is complete, there exists a vector $u \in V$ with $\|u\| = 1$ such that $\|u_n - u\| \to 0$. If x is the corresponding element of $P(V)$, it follows directly that $d(x_n,x) \to 0$.

Suppose next that V is finite-dimensional and let $\{x_n\}$ be any sequence in $P(V)$. Choose $u_n \in x_n$ with $\|u_n\| = 1$. Since the unit sphere of V is compact, by Proposition 19, there exist a subsequence $\{v_n\}$ of $\{u_n\}$ and a vector $v \in V$ with $\|v\| = 1$ such that $\|v_n - v\| \to 0$. Let y_n,y be the elements of $P(V)$ corresponding to v_n,v. For $n > p$ we have $\|v_n - v\| < 1$ and hence $\|v_n + v\| > 1$. Consequently

$$d(y_n,y) = \|v_n - v\| \to 0. \quad \square$$

In conjunction with Corollary 15, Propositions 19 and 20 show that any complete dense linear geometry X with $2 < \dim X < \infty$, or with $\dim X = 2$ and the Desargues property, can be given the structure of a metric space so that every closed set is compact.

5 NOTES

The long line is a fruitful source of counterexamples in topology; see Steen and Seebach (1978). For the roots of Proposition 3 in the investigations by Gauss on non-Euclidean geometry, see Section 12.6 of Coxeter (1989). Proposition 4 is proved in Doignon (1976).

Separation properties of two convex sets, and in particular the Hahn–Banach theorem, play a fundamental role in functional analysis; see, for example, Bourbaki (1987) and Holmes (1975). The separation of several convex sets, which is important in optimization, is treated in Bair and Fourneau (1980) and Boltyanskii (1975).

Theorem 16 is contained in Doignon (1976), but naturally it had many forerunners. The theorem throws an interesting light on Examples II.1, VI.2 and VI.3. These examples can be given a common formulation in the following way: let X be a connected finite-dimensional Riemannian manifold such that any two points a,b of X are joined by a unique geodesic arc, and define the segment $[a,b]$ to be the set of all points on this arc. A theorem of Beltrami, in conjunction with Theorem 16, shows that the three cited examples are the only complete unending linear geometries which can be obtained in this way if $\dim X > 2$, or if $\dim X = 2$ and X has the Desargues property. Beltrami's theorem states that if X is a connected Riemannian n-manifold such that every point has a neighbourhood which can be mapped homeomorphically into $\mathbb{R}^n$, so that geodesic arcs are mapped onto straight line segments, then X has constant curvature. The theorem is proved in Spivak (1975), for example. A proof which assumes only continuity of the Riemannian metric is sketched by Pogorelov (1991).

If $\dim X = 2$ and X does not have the Desargues property, then the possibilities are by no means so restricted. Busemann (1955), Section 11, shows that in Example VI.5 a metric d can be defined on $\mathbb{R}^2$ so that, for any points $x,y,z \in \mathbb{R}^2$,

(†) $$d(x,y) = d(x,z) + d(z,y) \text{ if and only if } z \in [x,y].$$

It follows that $[x,y]$ is the unique geodesic arc joining x and y. Another proof, using methods derived from integral geometry, is described in Salzmann (1967). It may be asked: *can a metric* d *with the property* (†) *be defined on any complete unending two-dimensional linear geometry?*

For an approach which starts from a metric space (X,d) and defines the segment $[x,y]$ to be the set of all points z such that $\mathrm{d}(x,y) = \mathrm{d}(x,z) + \mathrm{d}(z,y)$, see Rinow (1961) (and p. 420, in particular, for the connection with our approach).

Proposition 18 was first proved by Szczerba (1972) for two-dimensional V, using model theory rather than filters. The main ingredients in the proof of Lemma 17, namely the fundamental theorem of algebra and the fundamental theorem of Galois theory, are proved in Lang (1984), for example. Lemma 17 is actually a special case of a more general result of Artin and Schreier: if an algebraically closed field F has a proper subfield K of which it is a finite extension, then F has characteristic zero and $F = K(\mathrm{i})$, where $\mathrm{i}^2 = -1$. See, for example, Priess-Crampe (1983).

IX
Spaces of Convex Sets

This chapter is concerned with *spaces* of convex sets. We show that a metric may be defined on the space $\mathcal{F}(X)$ of all nonempty bounded closed subsets of a metric space X. Furthermore this *Hausdorff metric* is complete if the metric of X is complete. These results are then applied to the space $\mathcal{C}(X)$ of all nonempty bounded closed *convex* subsets of a *normed real vector space* X. The vector sum of two elements of $\mathcal{C}(X)$ need not again be in $\mathcal{C}(X)$, if X is infinite-dimensional. However, the closure of the vector sum is in $\mathcal{C}(X)$ and, with this modified definition of addition, $\mathcal{C}(X)$ has the structure of a commutative semigroup. Rådström's cancellation law implies that this semigroup may be embedded in a group. In fact $\mathcal{C}(X)$, with the Hausdorff metric and with the partial order induced by inclusion, may be embedded isometrically and isomorphically in a vector lattice with two further properties, which guarantee the existence of a norm. Vector lattices with these two further properties are here called *Kakutani spaces*.

A basic theorem, the Krein–Kakutani theorem, says that an arbitrary Kakutani space can be mapped isometrically and isomorphically onto a dense subset of the Kakutani space $C(K)$ of all continuous real-valued functions on some compact Hausdorff space K. Thus, as a final result, for a given normed real vector space X there exists a compact Hausdorff space K such that the space $\mathcal{C}(X)$ of all nonempty bounded closed convex subsets of X may be embedded in the space $C(K)$ of all continuous real-valued functions on K. To make the discussion self-contained, we also prove the Krein–Kakutani theorem.

1 THE HAUSDORFF METRIC

Let (X,d) be a metric space and let $\mathcal{F}(X)$ denote the family of all nonempty bounded closed subsets of X. We are going to show that a metric can be defined also on $\mathcal{F}(X)$.

We first define the *deviation* of a point $y \in X$ from a set $A \in \mathcal{F}(X)$ by

$$\delta(y,A) = \inf_{x \in A} d(x,y).$$

It is obvious that $\delta(y,A) = 0$ if $y \in A$. On the other hand $\delta(y,A) > 0$ if $y \notin A$. For, since the complement of A is open, there exists $\rho > 0$ such that no point of A is contained in the open ball with centre y and radius ρ.

The *Hausdorff distance* between two sets $A,B \in \mathcal{F}(X)$ is now defined by

$$h(A,B) = \max \{\sup_{y \in B} \delta(y,A), \sup_{x \in A} \delta(x,B)\}.$$

Evidently $h(A,B) < \infty$, since A and B are bounded. Moreover $h(A,B) \geq 0$, with equality if and only if $A = B$. It is obvious from the definition that $h(A,B) = h(B,A)$. It remains to prove the triangle inequality

$$h(A,B) \leq h(A,C) + h(C,B).$$

For all $x \in A$, $y \in B$, $z \in C$ we have

$$d(x,y) \leq d(x,z) + d(z,y).$$

Hence

$$\begin{aligned} \delta(x,B) &\leq d(x,z) + \delta(z,B) \\ &\leq d(x,z) + \sup_{w \in C} \delta(w,B) \\ &\leq d(x,z) + h(C,B). \end{aligned}$$

Since this inequality holds for all $z \in C$, it follows that

$$\begin{aligned} \delta(x,B) &\leq \delta(x,C) + h(C,B) \\ &\leq \sup_{u \in A} \delta(u,C) + h(C,B) \\ &\leq h(A,C) + h(C,B). \end{aligned}$$

Thus

$$\sup_{x \in A} \delta(x,B) \leq h(A,C) + h(C,B).$$

Similarly we can show that

$$\sup_{y \in B} \delta(y,A) \leq h(A,C) + h(C,B).$$

Taking both together, we obtain the triangle inequality.

From now on we understand the space $\mathcal{F}(X)$ to be equipped with the Hausdorff metric. Thus we can talk about convergent sequences, and Cauchy sequences, of nonempty bounded closed subsets of X.

PROPOSITION 1 *If* (X,d) *is a complete metric space, then* ($\mathcal{F}$(X),h) *is also a complete metric space.*

Proof We wish to show that if $\{A_n\}$ is any Cauchy sequence in $\mathcal{F}(X)$, then there exists $A \in \mathcal{F}(X)$ such that $A_n \to A$ as $n \to \infty$. For each $n = 1,2, \ldots$ let F_n denote the closure of the set $\bigcup_{m \geq n} A_m$, so that $F_1 \supseteq F_2 \supseteq \cdots$. Then $A = \bigcap_{n=1}^{\infty} F_n$ is a bounded closed subset of X. We will show that A has the required properties.

Fix any $\varepsilon > 0$. For each $k = 1,2, \cdots$ there exists a positive integer n_k such that

$$h(A_m,A_n) < \varepsilon/2^{k+1} \quad \text{for } m,n \geq n_k.$$

Moreover we may assume that $n_1 < n_2 < \cdots$. Choose any point $x_1 \in A_{n_1}$ and define inductively a sequence $\{x_k\}$ so that $x_k \in A_{n_k}$ and $d(x_k,x_{k-1}) < \varepsilon/2^k$. Then $\{x_k\}$ is a Cauchy sequence, since $d(x_j,x_k) < \varepsilon/2^k$ for all $j > k$. Since X is complete, there exists a point $y_1 \in X$ such that $x_k \to y_1$ as $k \to \infty$. Clearly, $d(x_1,y_1) \leq \varepsilon$. Moreover $y_1 \in A$, since $x_j \in F_{n_k}$ for all $j > k$ and F_{n_k} is closed. Thus A is nonempty and

$$\delta(x,A) \leq \varepsilon \quad \text{for all } x \in A_{n_1}.$$

On the other hand, let B denote the set of all points $y \in X$ such that $\delta(y,A_{n_1}) \leq \varepsilon$. Then $A_n \subseteq B$ for all $n \geq n_1$. Since B is closed, it follows that $F_{n_1} \subseteq B$ and hence $A \subseteq B$. Thus

$$\delta(y,A_{n_1}) \leq \varepsilon \quad \text{for all } y \in A.$$

Therefore $h(A,A_{n_1}) \leq \varepsilon$ and, by the triangle inequality, $h(A,A_n) < 2\varepsilon$ for all $n \geq n_1$. Thus we have shown that $A_n \to A$ as $n \to \infty$. □

Let $\mathcal{K}(X)$ denote the family of all nonempty compact subsets of X. Then $\mathcal{K}(X) \subseteq \mathcal{F}(X)$, since any compact subset of X is closed and bounded.

PROPOSITION 2 *If* (X,d) *is a complete metric space, then* ($\mathcal{K}$(X),h) *is also a complete metric space.*

Proof We need only show that $\mathcal{K}(X)$ is a closed subset of the complete metric space ($\mathcal{F}(X)$,h). Suppose $A_n \in \mathcal{K}(X)$ and $A_n \to A$ in $\mathcal{F}(X)$. We wish to show that $A \in \mathcal{K}(X)$. Thus we wish to show that any sequence $\{x_n\}$ of elements of A has a subsequence converging to an element of A. In fact, since X is complete and A is closed, it is enough to show that the sequence $\{x_n\}$ has a subsequence which is a Cauchy sequence.

There exists a positive integer n_1 such that $h(A,A_n) \leq 1/2$ for all $n \geq n_1$. Since A_{n_1} is compact, it is covered by finitely many open balls with radius 1/2. It follows that A is covered by finitely many open balls with the same centres and radius 1. Thus there is a ball B_1 with radius 1 which contains x_n for infinitely many n. In the same way there is a ball B_2 with radius 1/2 such that $B_1 \cap B_2$ contains x_n for infinitely many n. In general, for each positive integer k there is a ball B_k with radius $1/2^{k-1}$ such that $B_1 \cap ... \cap B_k$ contains x_n for infinitely many n. Hence there exists an increasing sequence $\{n_k\}$ of positive integers such that $x_{n_k} \in B_1 \cap ... \cap B_k$. Evidently the subsequence $\{x_{n_k}\}$ is a Cauchy sequence. □

The next result shows that the space of nonempty compact subsets of X has the Bolzano–Weierstrass property if X itself has. The proof uses the diagonal process which Cantor first used to show that the set of real numbers is uncountable.

PROPOSITION 3 *If* (X,d) *is a compact metric space, then* $(\mathcal{K}(X),h)$ *is also a compact metric space.*

Proof Since a compact metric space is necessarily complete, $(\mathcal{K}(X),h)$ is a complete metric space by Proposition 2. Thus we need only show that any sequence $\{A_n\}$ of elements of $\mathcal{K}(X)$ has a subsequence which is a Cauchy sequence.

For a given $\varepsilon_1 > 0$ there is a positive integer m_1 such that the space X is covered by finitely many open balls $B_{11},...,B_{1m_1}$ of radius ε_1. With each set A_n we associate the collection of those balls B_{1j} with which it has at least one point in common. Since a set with m_1 elements has just $2^{m_1} - 1$ nonempty subsets, there exists a collection of balls B_{1j} which is associated in this way with infinitely many sets A_n. Let $\{A_{n,1}\}$ be a subsequence of $\{A_n\}$ such that with each set $A_{n,1}$ there is associated the same collection of balls B_{1j}. Then $\delta(x,A_{n,1}) < 2\varepsilon_1$ if $x \in A_{m,1}$ and hence $h(A_{m,1},A_{n,1}) \leq 2\varepsilon_1$ for all m,n.

We now repeat this process with ε_1 replaced by $\varepsilon_2 = \varepsilon_1/2$ and $\{A_n\}$ replaced by $\{A_{n,1}\}$. We obtain a subsequence $\{A_{n,2}\}$ of $\{A_{n,1}\}$ such that $h(A_{m,2},A_{n,2}) \leq 2\varepsilon_2 = \varepsilon_1$ for all m,n. By repeating the process infinitely often and constructing the diagonal sequence $\{A_n'\}$, where $A_n' = A_{n,n}$, we obtain a subsequence of $\{A_n\}$ which is a Cauchy sequence. □

2 THE SPACE $\mathscr{C}(X)$

To apply the preceding results to convex sets we now impose more structure on the metric space. *Throughout this section we assume that X is a normed real vector space.* Thus with each $x \in X$ there is associated a real number $\|x\|$ with the properties

(N1) $\|x\| \geq 0$, *with equality if and only if* $x = 0$,
(N2) $\|x + y\| \leq \|x\| + \|y\|$ *for all* $x,y \in X$,
(N3) $\|\alpha x\| = |\alpha| \, \|x\|$ *for all* $x \in X$ *and all* $\alpha \in \mathbb{R}$,

and X is a metric space with the metric $d(x,y) = \|x - y\|$. We will denote by U the *closed unit ball* of X: $U = \{x \in X: \|x\| \leq 1\}$. Evidently U is convex, by **(N1)–(N3)**.

Let $\mathscr{C} = \mathscr{C}(X)$ denote the collection of all nonempty bounded closed convex subsets of X. If $A,B \in \mathscr{C}$, then $A + B$ is a nonempty bounded convex subset of X, but it need not be closed if X is infinite-dimensional. However, if we define $A \oplus B$ to be the closure of $A + B$, then $A \oplus B \in \mathscr{C}$.

The properties in Chapter VII, Section 3, about addition of convex sets in a vector space over an ordered division ring, remain valid in $\mathscr{C} = \mathscr{C}(X)$ with this new definition of addition:

(i) $A \oplus B = B \oplus A$, $(A \oplus B) \oplus C = A \oplus (B \oplus C)$,
(ii) $\alpha(A \oplus B) = \alpha A \oplus \alpha B$, $(\alpha\beta)A = \alpha(\beta A)$, $1A = A$,
(iii) $A \oplus 0 = A$, $0A = 0$,
(iv) *if* $A \subseteq B$, *then* $A \oplus C \subseteq B \oplus C$ *and* $\alpha A \subseteq \alpha B$,
(v) *if* $\lambda,\mu \geq 0$, *then* $(\lambda + \mu)A = \lambda A \oplus \mu A$.

Since the proofs for the other properties are similar but simpler, we prove only the associativity of addition:

$$(A \oplus B) \oplus C = A \oplus (B \oplus C).$$

Let D denote the closure of the set $\{a + b + c: a \in A, b \in B, c \in C\}$. If $x \in A \oplus B$ then, for each positive integer n, there exist $a_n \in A$, $b_n \in B$ and $u_n \in U$ such that $x = u_n/n + a_n + b_n$. Hence if $c \in C$, then $x + c \in D$. Thus $(A \oplus B) + C \subseteq D$ and hence, since D is closed, $(A \oplus B) \oplus C \subseteq D$. Since the reverse inequality is obvious, this proves that $(A \oplus B) \oplus C = D$. But in the same way we can prove that $A \oplus (B \oplus C) = D$.

We now derive some additional properties. If $\mathscr{C}$ is partially ordered by inclusion, then

(vi) *any* $A,B \in \mathscr{C}$ *have a least upper bound* $A \vee B$.

In fact $A \vee B$ is the closure of $[A \cup B]$. Furthermore,

(vii) *for any* $A,B,C \in \mathscr{C}$, $(A \vee B) \oplus C = (A \oplus C) \vee (B \oplus C)$.

To prove this we need only show that $(A \vee B) \oplus C \subseteq (A \oplus C) \vee (B \oplus C)$, since the reverse inequality is obvious. A similar argument to that used in proving the associativity of addition shows that $(A \vee B) \oplus C$ is the closure of the set $[A \cup B] + C$. But if $x \in [A \cup B] + C$ then, for some $a \in A$, $b \in B$, $c \in C$ and $\lambda \in [0,1]$,

$$x = \lambda a + (1 - \lambda)b + c = \lambda(a + c) + (1 - \lambda)(b + c).$$

Thus $x \in (A \oplus C) \vee (B \oplus C)$, and hence $(A \vee B) \oplus C \subseteq (A \oplus C) \vee (B \oplus C)$.

Of particular importance is the following *cancellation law* for addition:

PROPOSITION 4 *For any* $A,B,C \in \mathscr{C}$,

(viii) *if* $A \oplus C \subseteq B \oplus C$, *then* $A \subseteq B$.

Proof Choose $a \in A$ and $c_1 \in C$. Since $a + c_1 \in B \oplus C$, there exist $b_1 \in B$, $c_2 \in C$ and $x_1 \in 2^{-1}U$ such that

$$a + c_1 = b_1 + c_2 + x_1.$$

Repeating this procedure, we define inductively $b_k \in B$, $c_{k+1} \in C$ and $x_k \in 2^{-k}U$ such that

$$a + c_k = b_k + c_{k+1} + x_k \quad (k \geq 1).$$

Adding these relations for $k = 1,\dots,n$ and mutiplying by $1/n$, we obtain

$$a + c_1/n = c_{n+1}/n + b_n' + x_n',$$

where $b_n' = (b_1 + \dots + b_n)/n \in B$, since B is convex, and $x_n' = (x_1 + \dots + x_n)/n \in (1/n)U$, by construction. Since C is bounded and B is closed, it follows that $a \in B$. Thus $A \subseteq B$. □

COROLLARY 5 *If $A,B,C \in \mathscr{C}$ and $A \oplus C = B \oplus C$, then $A = B$.* □

To illustrate the role played by convexity in this result we now show that if $X = \mathbb{R}^d$, $A \subseteq X$ and $C = [A]$, then

$$d^{-1}A + C = d^{-1}C + C.$$

Since the left side is obviously contained in the right, and since $d^{-1}C + C = d^{-1}(d+1)C$, we need only show that $C \subseteq (d+1)^{-1}A + d(d+1)^{-1}C$. If $x \in C$ then, by Corollary III.18, there exists a finite affine independent set $F \subseteq A$ such that $x \in [F]$. Thus $x = \sum_{i=1}^{m} \lambda_i x_i$, where $x_i \in A$ and $\lambda_i > 0$ $(i = 1,\ldots,m)$, $\sum_{i=1}^{m} \lambda_i = 1$ and $m \leq d+1$. Moreover we may choose the notation so that $\lambda_1 = \max_i \lambda_i$. Then $\lambda_1 \geq m^{-1} \geq (d+1)^{-1}$ and

$$\begin{aligned} x &= (d+1)^{-1}x_1 + [\lambda_1 - (d+1)^{-1}]x_1 + \textstyle\sum_{i=2}^{m} \lambda_i x_i \\ &= (d+1)^{-1}x_1 + d(d+1)^{-1} \textstyle\sum_{i=1}^{m} \mu_i x_i, \end{aligned}$$

where $\mu_i \geq 0$ $(i = 1,\ldots,m)$ and $\sum_{i=1}^{m} \mu_i = 1$. Thus

$$x \in (d+1)^{-1}A + d(d+1)^{-1}C.$$

If we choose A bounded and closed, but not convex, then C is a bounded closed convex set, but $d^{-1}A \neq d^{-1}C$.

For a normed vector space the definition of the Hausdorff metric can be reformulated in the following way:

LEMMA 6 *If X is a normed real vector space then, for any sets $A,B \in \mathscr{F}(X)$,*

$$\mathrm{h}(A,B) = \inf\{\lambda \geq 0 : A \subseteq B \oplus \lambda U, B \subseteq A \oplus \lambda U\}.$$

Proof If $A \subseteq B \oplus \lambda U$ then, for any $x \in A$, $\delta(x,B) \leq \lambda$ and consequently $\sup_{x \in A} \delta(x,B) \leq \lambda$. Hence if $A \subseteq B \oplus \lambda U$ and $B \subseteq A \oplus \lambda U$, then $\mathrm{h}(A,B) \leq \lambda$.

On the other hand if $A \not\subseteq B \oplus \mu U$ then, for some $x \in A$, $x \notin B \oplus \mu U$. Hence $\mathrm{d}(x,y) > \mu$ for all $y \in B$, and so $\delta(x,B) \geq \mu$. Consequently, if either $A \not\subseteq B \oplus \mu U$ or $B \not\subseteq A \oplus \mu U$, then $\mathrm{h}(A,B) \geq \mu$. The result follows. □

LEMMA 7 *If X is a normed real vector space, if $\{A_n\}$ is a sequence of sets in $\mathscr{C}(X)$ and if $A_n \to A$ in $(\mathscr{F}(X),\mathrm{h})$, then $A \in \mathscr{C}(X)$.*

Proof Since A is nonempty, bounded and closed, we need only show that it is convex. Let $z = \lambda x + (1-\lambda)y$, where $x,y \in A$ and $0 < \lambda < 1$. By Lemma 6, for any $\varepsilon > 0$ there is a positive integer m such that

$$A_n \subseteq A + \varepsilon U, \quad A \subseteq A_n + \varepsilon U \quad \text{for every } n \geq m.$$

Since A_n and U are convex, so also is $A_n + \varepsilon U$. Hence

$$z \in A_n + \varepsilon U \subseteq A + 2\varepsilon U.$$

Since ε is arbitrary and A is closed, it follows that $z \in A$. □

From Lemma 7 and Proposition 1 we immediately obtain

PROPOSITION 8 *If X is a complete normed real vector space then, with the Hausdorff metric* h *derived from the norm, $(\mathscr{C}(X),\mathrm{h})$ is also a complete metric space.* □

A finite-dimensional normed real vector space X is necessarily complete, and a subset of X is compact if and only if it is both bounded and closed. Hence, by applying Lemma 7 and Proposition 3 to a compact set $X' \subseteq X$ we immediately obtain the *Blaschke selection principle*:

PROPOSITION 9 *If X is a finite-dimensional normed real vector space then, with the Hausdorff metric* h *derived from the norm, any bounded sequence of sets in $(\mathscr{C}(X),\mathrm{h})$ has a convergent subsequence.* □

3 EMBEDDINGS OF $\mathscr{C}(X)$

Since $\mathscr{C}$ has the structure of a commutative semigroup under addition, with a zero element and with multiplication by non-negative real numbers as a semigroup of operators, and since the cancellation law for addition holds, it is possible to embed $\mathscr{C}$ in a vector space. We now sketch the construction, which is similar to that by which the semigroup of non-negative integers is embedded in the group of all integers (and to that used at the end of Section 1 in Chapter VIII).

If (A,B) and (C,D) are ordered pairs of elements of $\mathscr{C}$, we write $(A,B) \sim (C,D)$ if $A \oplus D = B \oplus C$. It follows directly from Corollary 5 that $\sim$ is an equivalence relation. Let $\{A,B\}$ denote the equivalence class containing the pair (A,B) and let $\mathscr{L} = \mathscr{L}(X)$ denote the set of all equivalence classes $\{A,B\}$.

If $\{A,B\}$ and $\{C,D\}$ are elements of $\mathscr{L}$, we define their sum by

$$\{A,B\} + \{C,D\} = \{A \oplus C, B \oplus D\}.$$

It is easily seen that the sum is uniquely defined, i.e. if $\{A',B'\} = \{A,B\}$ and $\{C',D'\} = \{C,D\}$ then

$$\{A',B'\} + \{C',D'\} = \{A,B\} + \{C,D\}.$$

Similarly, for any $\alpha \in \mathbb{R}$ we can define

$$\begin{aligned} \alpha\{A,B\} &= \{\alpha A, \alpha B\} && \text{if } \alpha \geq 0, \\ &= \{(-\alpha)B, (-\alpha)A\} && \text{if } \alpha < 0. \end{aligned}$$

It is not difficult to verify that with these definitions $\mathscr{L}$ is a *vector space* with $\{0,0\}$ as the zero vector. Evidently any $\{A,B\} \in \mathscr{L}$ has the form

$$\{A,B\} = \{A,0\} - \{B,0\}.$$

We can also define an order relation on $\mathscr{L}$ by

$$\{A,B\} \leq \{C,D\} \text{ if } A \oplus D \subseteq B \oplus C.$$

It follows from Proposition 4 that this order relation is uniquely defined and is indeed a partial order. Also, if $\{A,B\} \leq \{C,D\}$ then, for any $\{E,F\} \in \mathscr{L}$ and any $\lambda \geq 0$,

$$\begin{gathered} \{A,B\} + \{E,F\} \leq \{C,D\} + \{E,F\}, \\ \lambda\{A,B\} \leq \lambda\{C,D\}. \end{gathered}$$

Furthermore any two elements $\{A,B\}$ and $\{C,D\}$ of $\mathscr{L}$ have a least upper bound, namely $\{G, B \oplus D\}$, where $G = (A \oplus D) \vee (B \oplus C)$. For we certainly have

$$\begin{aligned} \{A,B\} &= \{A \oplus D, B \oplus D\} \leq \{G, B \oplus D\}, \\ \{C,D\} &= \{B \oplus C, B \oplus D\} \leq \{G, B \oplus D\}. \end{aligned}$$

On the other hand, if $\{A,B\} \leq \{E,F\}$ and $\{C,D\} \leq \{E,F\}$, then

$$A \oplus D \oplus F \subseteq B \oplus D \oplus E, \; B \oplus C \oplus F \subseteq B \oplus D \oplus E,$$

and hence

$$G \oplus F = (A \oplus D \oplus F) \vee (B \oplus C \oplus F) \subseteq B \oplus D \oplus E.$$

Thus $\mathscr{L}$ is a *vector lattice.*

We draw attention also to two other properties of $\mathscr{L}$. First, for any $\{A,B\} \in \mathscr{L}$ there is a positive integer n such that

$$-n\{U,0\} \leq \{A,B\} \leq n\{U,0\}.$$

Indeed we can choose the positive integer m so large that $A,B \subseteq mU$ and then, if $n \geq 2m$, we will have $A \subseteq B + nU$ and $B \subseteq A + nU = A + (-n)U$.

Secondly, if $\{A,B\},\{C,D\} \in \mathscr{L}$ and $n\{A,B\} \leq \{C,D\}$ for every positive integer n, then $\{A,B\} \leq \{0,0\}$. Equivalently, if $nA \oplus D \subseteq nB \oplus C$ for every positive integer n, then $A \subseteq B$. Indeed if $a \in A$ and $d \in D$, then there exist $b_n \in B$, $c_n \in C$ and $u_n \in U$ such that

$$na + d = nb_n + c_n + u_n.$$

Since C and D are bounded, and B is closed, it follows that $a \in B$.

In the terminology which will be introduced in the next section, we have shown that $\mathscr{L}$ is a *Kakutani space.* It is shown there that a Kakutani space may always be normed. Applying the definition of the norm to $\mathscr{L}$ we immediately verify that the distance between $\{A,0\}$ and $\{B,0\}$, with respect to this norm, is precisely the Hausdorff distance

$$\mathrm{h}(A,B) = \inf \{\lambda \geq 0: A \subseteq B \oplus \lambda U, B \subseteq A \oplus \lambda U\}.$$

If we define a map $\varphi: \mathscr{C} \to \mathscr{L}$ by $\varphi(A) = \{A,0\}$ for every $A \in \mathscr{C}$, then the preceding argument shows that

$$\varphi(A \oplus B) = \varphi(A) + \varphi(B), \quad \varphi(\lambda A) = \lambda\varphi(A) \quad \text{for any } \lambda \geq 0,$$
$$A \subseteq B \text{ if and only if } \varphi(A) \leq \varphi(B), \quad \varphi(A \vee B) = \varphi(A) \vee \varphi(B),$$
$$\mathrm{h}(A,B) = \|\varphi(A) - \varphi(B)\|.$$

Thus we have proved

THEOREM 10 *Let X be a normed real vector space and, with the Hausdorff metric* h *derived from the norm, let* $(\mathcal{C}(X),\mathrm{h})$ *be the metric space of nonempty bounded closed convex subsets of X.*

Then there exists an injective map φ *of* $\mathcal{C} = \mathcal{C}(X)$ *into a Kakutani space* $\mathcal{L}$ *with the following properties:*

(i) φ *preserves sums and non-negative real multiples,*

(ii) φ *preserves order and the supremum of any two elements,*

(iii) φ *preserves distance,*

(iv) $\mathcal{C}' = \varphi(\mathcal{C})$ *is a convex cone in* $\mathcal{L}$ *and* $\mathcal{C}'$ *generates* $\mathcal{L}$:

$$\lambda\mathcal{C}' + \mu\mathcal{C}' \subseteq \mathcal{C}' \text{ for all } \lambda,\mu \geq 0 \text{ and } \mathcal{L} = \mathcal{C}' - \mathcal{C}'. \quad \square$$

By identifying A with $\varphi(A)$, we may regard $\mathcal{C}$ as embedded in $\mathcal{L}$. From Theorem 10 we immediately obtain the *translation invariance* of the Hausdorff metric (which can also be proved directly):

COROLLARY 11 *Let X be a normed real vector space and, with the Hausdorff metric* h *derived from the norm, let* $(\mathcal{C}(X),\mathrm{h})$ *be the metric space of nonempty bounded closed convex subsets of X. If* $A,B,C \in \mathcal{C}(X)$ *and* $\lambda \geq 0$, *then*

$$\mathrm{h}(A \oplus C, B \oplus C) = \mathrm{h}(A,B), \quad \mathrm{h}(\lambda A, \lambda B) = \lambda \mathrm{h}(A,B). \quad \square$$

The prototype of a Kakutani space is the space $C(K)$ of all continuous functions $f: K \to \mathbb{R}$, where K is a compact Hausdorff space and where, if $f,g \in C(K)$ and $\alpha \in \mathbb{R}$, $f+g$ and αf are defined by

$$(f+g)(t) = f(t) + g(t), \quad (\alpha f)(t) = \alpha f(t) \quad \text{for every } t \in K,$$

and $f \leq g$ is defined by

$$f(t) \leq g(t) \quad \text{for every } t \in K.$$

Then any $f,g \in C(K)$ have a least upper bound $f \vee g$ defined by

$$(f \vee g)(t) = \max\{f(t), g(t)\} \quad \text{for every } t \in K.$$

Moreover, for any $f,g \in C(K)$, if $nf \leq g$ for every positive integer n, then $f \leq 0$. Finally, for any $f \in C(K)$ there is a positive integer n such that $-ne \leq f \leq ne$, where $e \in C(K)$ is the constant function defined by

$$e(t) = 1 \quad \text{for every } t \in K.$$

The associated norm for the Kakutani space $C(K)$ is the supremum norm:

$$\|f\| = \sup \{|f(t)|: t \in K\}.$$

The Krein–Kakutani theorem, which will be proved in the next section, says that an arbitrary Kakutani space can be mapped isomorphically and isometrically onto a dense subset of the Kakutani space $C(K)$ for some compact Hausdorff space K. By composing the injective map $\varphi: \mathscr{C}(X) \to \mathscr{L}$ of Theorem 10, with the linear isometry and lattice isomorphism J of $\mathscr{L}$ into $C(K)$ guaranteed by the Krein–Kakutani theorem, we immediately obtain

THEOREM 12 *Let X be a normed real vector space and, with the Hausdorff metric* h *derived from the norm, let* $(\mathscr{C}(X),\mathrm{h})$ *be the metric space of nonempty bounded closed convex subsets of X.*

Then there exist a compact Hausdorff space K and an injective map ψ of $\mathscr{C} = \mathscr{C}(X)$ into the Kakutani space $C(K)$ *of all continuous real-valued functions on K with the following properties:*

(i) *ψ preserves sums and non-negative real multiples,*
(ii) *ψ preserves order and the supremum of any two elements,*
(iii) *ψ preserves distance,*
(iv) *$\psi(U)$ is the constant function* 1,
(v) *$\mathscr{C}' = \psi(\mathscr{C})$ is a convex cone in* $C(K)$ *and $\mathscr{C}' - \mathscr{C}'$ is dense in* $C(K)$. □

By identifying A with $\psi(A)$, we may regard $\mathscr{C}$ as embedded in $C(K)$.

4 THE KREIN-KAKUTANI THEOREM

In this section we prove the Krein–Kakutani theorem, not only to make our account self-contained but also because the existence in our case of an order unit makes possible some simplifications. At the same time the proof may serve as a minicourse on convexity in functional analysis.

A *linear functional* on a normed real vector space V is a map $f: V \to \mathbb{R}$ such that

$$f(\alpha x + \beta y) = \alpha f(x) + \beta f(y) \quad \text{for all } x,y \in V \text{ and all } \alpha,\beta \in \mathbb{R}.$$

The linear functional f is said to be *bounded* if there exists a non-negative real number μ such that

$$|f(x)| \leq \mu\|x\| \text{ for all } x \in V.$$

The set V' of all bounded linear functionals on V is also a normed vector space if we define

$$(\alpha f + \beta g)(x) = \alpha f(x) + \beta g(x),$$
$$\|f\| = \sup\{|f(x)|: \|x\| = 1\}.$$

Since f is linear, we can also write

$$\|f\| = \sup\{|f(x)|: \|x\| \leq 1\}.$$

For any non-zero vector $x_0 \in V$, there is a linear functional $f \in V'$ such that $\|f\| = 1$ and $f(x_0) = \|x_0\|$. This is a special case, with $p(x) = \|x\|$, $W = \{\alpha x_0: \alpha \in \mathbb{R}\}$ and $g(\alpha x_0) = \alpha\|x_0\|$ for $\alpha \in \mathbb{R}$, of the analytic form of the Hahn–Banach theorem:

(HB) *Let W be a vector subspace of the real vector space V and let $p: V \to \mathbb{R}$ be a sublinear functional on V*, i.e.

$$p(x + y) \leq p(x) + p(y), \quad p(\lambda x) = \lambda p(x) \text{ for all } x,y \in V \text{ and all real } \lambda \geq 0.$$

If g is a linear functional on W such that

$$g(y) \leq p(y) \text{ for every } y \in W,$$

then there exists a linear functional f on V such that $f(y) = g(y)$ for every $y \in W$ and

$$f(x) \leq p(x) \text{ for every } x \in V.$$

Proof Let

$$A = \{(x,\alpha) \in V \times \mathbb{R}: p(x) \leq \alpha\}, \quad B = \{(x,\alpha) \in W \times \mathbb{R}: g(x) = \alpha\}.$$

Then A is a convex subset of $V \times \mathbb{R}$ and B is a vector subspace of $V \times \mathbb{R}$. We will show that $A^i = \{(x,\alpha) \in V \times \mathbb{R}: p(x) < \alpha\}$, which implies $A^i \cap B = \emptyset$.

Suppose $(x,\alpha) \in A^i$. If $\beta > 0$ then $(0,\beta) \in A$ and hence, for some $\lambda > 1$, $\lambda(x,\alpha) + (1 - \lambda)(0,\beta) \in A$. Thus

$$\lambda p(x) = p(\lambda x) \leq \lambda\alpha + (1 - \lambda)\beta,$$

which implies $p(x) < \alpha$. On the other hand, suppose $p(x) < \alpha$ and choose α' so that $p(x) < \alpha' < \alpha$. For any $(y,\beta) \in V \times \mathbb{R}$, if $0 < \theta < 1$ and θ is sufficiently close to 1, then

$$p(\theta x + (1-\theta)y) \leq \theta\alpha' + (1-\theta)p(y) < \theta\alpha + (1-\theta)\beta.$$

Hence $(x,\alpha) \in A^i$, by Proposition V.27.

It follows from Proposition VIII.11 that there exists a hyperplane H of $V \times \mathbb{R}$ such that $B \subseteq H$ and $A^i \cap H = \emptyset$. Since any element of $V \times \mathbb{R}$ can be uniquely expressed as the vector sum of an element of H and a scalar multiple of a fixed element of A^i, there exists a linear functional h on $V \times \mathbb{R}$ such that $h(x,\alpha) = 0$ for $(x,\alpha) \in H$ and $h(x,\alpha) > 0$ for $(x,\alpha) \in A^i$. For $y \in W$ we have

$$h(y,g(y)) = 0 < h(y,1+p(y))$$

and hence

$$0 < h(0,1+p(y)-g(y)) = [1+p(y)-g(y)]h(0,1).$$

Since $g(y) \leq p(y)$, it follows that $h(0,1) > 0$. For any $(x,\alpha) \in V \times \mathbb{R}$,

$$h(x,\alpha) = h(x,0) + \alpha h(0,1).$$

Hence $h(x,\alpha) = 0$ if and only if $\alpha = f(x) := -h(x,0)/h(0,1)$. Evidently f is a linear functional on V and $f(y) = g(y)$ for $y \in W$. Furthermore $f(x) \leq p(x)$ for any $x \in V$, since

$$0 \leq h(x,p(x)) = [p(x)-f(x)]h(0,1). \quad \square$$

Since V' is also a normed vector space, we can in the same way define the normed vector space $V'' = (V')'$ of all bounded linear functionals on V'. For any $x \in V$, the map J_x: $V' \to \mathbb{R}$ defined by $J_x(f) = f(x)$ is a bounded linear functional on V'. Moreover $\|J_x\| = \|x\|$, since

$$\|J_x\| = \sup\{|J_x(f)|: \|f\| = 1\} = \sup\{|f(x)|: \|f\| = 1\}$$

and since, for any $x \neq 0$, there exists $f \in V'$ with $\|f\| = 1$ and $f(x) = \|x\|$. It follows from (HB) also that the map $x \to J_x$ is injective. Since $J_{\alpha x+\beta y} = \alpha J_x + \beta J_y$, the map $x \to J_x$ is actually an isometric isomorphism of V into V''.

Since V' is a normed vector space, it has a topology derived from the metric $d(f,g) = \|f-g\|$. However, there is a weaker topology on V' which is more convenient for our purposes. The open sets of this *weak* topology* are the unions of sets of the form

$$N(f_0,\varepsilon,A) = \{f \in V': |f(x) - f_0(x)| < \varepsilon \text{ for all } x \in A\},$$

where $f_0 \in V'$, $\varepsilon > 0$ and A is a finite subset of V. The weak* topology is a Hausdorff topology, since if f,g are distinct elements of V', then $f(x) \neq g(x)$ for some $x \in V$ and hence there exist disjoint neighbourhoods $N(f,\varepsilon,x)$, $N(g,\varepsilon,x)$.

The convenience of the weak* topology stems from the Banach–Alaoglu theorem:

(BA) *The closed unit ball* $U' = \{f \in V': \|f\| \leq 1\}$ *of* V' *is compact in the weak* topology.*

Proof If $f \in U'$, then $|f(x)| \leq \|x\|$ and so $f(x)$ lies in the compact interval $I_x = [-\|x\|,\|x\|]$. If $P = \prod_{x \in V} I_x$ then $U' \subseteq P$, since P is the set of all functions $g: V \to \mathbb{R}$ such that $g(x) \in I_x$ for every $x \in V$. Moreover the topology U' inherits as a subspace of P is the weak* topology. But P is a product of compact spaces and hence itself compact, by Tychonoff's theorem. Hence, to show that U' is compact, we need only show that it is a closed subset of P.

Let $g \in P$ be a point in the closure of U'. Then g maps V into $\mathbb{R}$ and $|g(x)| \leq \|x\|$ for each $x \in V$. Suppose $z = \alpha x + \beta y$, where $x,y \in V$ and $\alpha,\beta \in \mathbb{R}$. For each $\varepsilon > 0$, the set $N = \{h \in P: |h(t) - g(t)| < \varepsilon \text{ for all } t \in A\}$, where $A = \{x,y,z\}$, is an open subset of P which contains g. Since g is in the closure of U', there exists $f \in U' \cap N$, and since f is linear, $f(z) = \alpha f(x) + \beta f(y)$. Hence

$$|g(z) - \alpha g(x) - \beta g(y)| < (1 + |\alpha| + |\beta|)\varepsilon.$$

Since this inequality holds for each $\varepsilon > 0$, we must actually have

$$g(z) = \alpha g(x) + \beta g(y).$$

Thus g is a linear functional, and moreover $g \in U'$. □

Another result which will be required is the Krein–Milman theorem:

(KM) *If C is a nonempty weak* compact convex subset of V', then the set K of all extreme points of C is nonempty and C is the weak* closure of the convex hull of K.*

Proof The family $\mathcal{F}$ of all nonempty weak* compact faces of C is nonempty, since it contains C itself. If we regard $\mathcal{F}$ as partially ordered by inclusion then, for any

$F \in \mathcal{F}$, there is a maximal totally ordered subfamily $\mathcal{T}$ containing F. The intersection E of all members of $\mathcal{T}$ is a face of C. Moreover E is nonempty and weak* compact, since each member of $\mathcal{T}$ is weak* compact. Thus $E \in \mathcal{F}$ and no proper subset of E is in $\mathcal{F}$. We will show that E is necessarily a singleton.

Assume on the contrary that E contains distinct points f,g of V'. Then $f(x) \neq g(x)$ for some $x \in V$. Put

$$\mu = \sup \{h(x): h \in E\},$$

and let D be the subset of all points of E at which the supremum is attained. Since E is weak* compact, D is nonempty and weak* closed. Hence D is also weak* compact. Moreover D is convex, since if $h_1,h_2 \in D$ and $h = \theta h_1 + (1 - \theta)h_2$, where $0 < \theta < 1$, then

$$h(x) = \theta\mu + (1 - \theta)\mu = \mu.$$

On the other hand, if $h \in D$ and $h = \theta h_1 + (1 - \theta)h_2$, where $h_1,h_2 \in E$ and $0 < \theta < 1$, then $h_1,h_2 \in D$, since $\mu = \theta h_1(x) + (1 - \theta)h_2(x)$ and $h_1(x) \leq \mu$, $h_2(x) \leq \mu$. Consequently D is a face of E, and therefore of C. But D is a proper subset of E, since it does not contain both f and g. Thus we have a contradiction.

This proves that any nonempty weak* compact face of C contains an extreme point of C. If ϕ is a bounded linear functional on V', a similar argument shows that the set of all points of C at which ϕ attains its supremum is a nonempty weak* compact face of C. Consequently the supremum is attained, in particular, at an extreme point of C.

Now let A be the weak* closure of the convex hull of the set K of all extreme points of C. Then $A \subseteq C$ and hence A is weak* compact. Assume there exists a point $f \in C \backslash A$. By applying Proposition VIII.8 to the sets $\{f\} + 2\rho U'$ and A for some small $\rho > 0$, we see that there exists a hyperplane of V' such that $\{f\} + \rho U'$ is contained in one of the open half-spaces associated with this hyperplane and A in the other. Thus, as in the proof of (HB), there are a linear functional ϕ on V' and a real number γ such that $\phi(f + \rho u') > \gamma$ for every $u' \in U'$ and $\phi(g) < \gamma$ for every $g \in A$. Since this implies

$$[\gamma - \phi(f)]/\rho < \phi(u') < -[\gamma - \phi(f)]/\rho \quad \text{for all } u' \in U',$$

the linear functional ϕ is actually bounded. Hence ϕ attains its supremum on C at an extreme point e of C. Since $e \in A$, $\phi(e) < \gamma < \phi(f)$. Hence $f \in C$, which is a contradiction. □

A real vector space V is said to be a *vector lattice* if a partial order $\leq$ is defined on V with the properties

(i) *if* $x,y \in V$ *and* $x \leq y$, *then* $x + z \leq y + z$ *for every* $z \in V$ *and* $\lambda x \leq \lambda y$ *for every real* $\lambda \geq 0$,
(ii) *any two vectors* $x,y \in V$ *have a least upper bound* $x \vee y$.

It follows that any two vectors $x,y \in V$ also have a greatest lower bound $x \wedge y$, namely

$$x \wedge y = -((-x) \vee (-y)).$$

We also have the following simple properties: *if* $x,y,z \in V$ *and* $\lambda \geq 0$, *then*

$$(x \vee y) + z = (x + z) \vee (y + z), \quad (x \wedge y) + z = (x + z) \wedge (y + z),$$
$$\lambda(x \vee y) = (\lambda x) \vee (\lambda y), \quad \lambda(x \wedge y) = (\lambda x) \wedge (\lambda y) \;\; \text{if } \lambda \geq 0,$$
$$(x \vee y) + (x \wedge y) = x + y.$$

[*Proof* Put

$$u = x \vee y, \quad v = (x + z) \vee (y + z), \quad w = (\lambda x) \vee (\lambda y).$$

From $x,y \leq u$ we obtain $x + z \leq u + z$, $y + z \leq u + z$ and hence $v \leq u + z$. On the other hand from $x + z \leq v$, $y + z \leq v$ we obtain $x \leq v - z$, $y \leq v - z$. Hence $u \leq v - z$ and $u + z \leq v$. This proves that $v = u + z$, and similarly we can prove that $w = \lambda u$ if $\lambda > 0$.

The corresponding relations with $\vee$ replaced by $\wedge$ follow immediately. Also, by what we first proved,

$$-(x \wedge y) + x = ((-x) \vee (-y)) + x = 0 \vee (x - y)$$

and hence

$$-(x \wedge y) + x + y = 0 \vee (x - y) + y = y \vee x. \quad \square]$$

These relations imply the lattice distributive laws: *if* $x,y,z \in V$, *then*

$$(x \vee y) \wedge z = (x \wedge z) \vee (y \wedge z), \quad (x \wedge y) \vee z = (x \vee z) \wedge (y \vee z).$$

[*Proof* Put

$$u = (x \wedge z) \vee (y \wedge z), \quad v = x \vee y.$$

Since $x \wedge z \leq v \wedge z$ and $y \wedge z \leq v \wedge z$, we certainly have $u \leq v \wedge z$. On the other hand, since

$$x \wedge z = x + z - (x \vee z), \quad y \wedge z = y + z - (y \vee z),$$

we have

$$x + z - (x \vee z) \leq u, \quad y + z - (y \vee z) \leq u,$$

and hence $x + z \leq u + (v \vee z)$, $y + z \leq u + (v \vee z)$. It follows that $v + z \leq u + (v \vee z)$, i.e. $v \wedge z \leq u$. This proves the first distributive law, and the second is a consequence. □]

In particular, if we put

$$x^+ = x \vee 0, \quad x^- = -(x \wedge 0) = (-x) \vee 0,$$

then

$$x^+ \geq 0, x^- \geq 0, x^+ \wedge x^- = 0,$$
$$x^+ - x^- = x, \quad x^+ + x^- = x \vee (-x).$$

For convenience of writing we put

$$V_+ = \{x \in V : x \geq 0\}.$$

We will say that a vector lattice V is a *Kakutani space* if it has the additional properties

(iii) *if $x,y \in V$ and $nx \leq y$ for every positive integer n, then $x \leq 0$,*
(iv) *there exists $e \in V$ such that, for each $x \in V$, there is a positive integer n for which $-ne \leq x \leq ne$.*

For any $x \in V$, put

$$\|x\| = \inf \{\lambda > 0 : -\lambda e \leq x \leq \lambda e\}.$$

It is easily verified that with this definition the Kakutani space V is a normed vector space. Moreover norm and order are connected by the properties

(N4) $\|x\| = \|x^+ + x^-\|$ *for every* $x \in V$,
(N5) $\|x \vee y\| = \max\{\|x\|, \|y\|\}$ *if* $x,y \in V_+$.

The real line $\mathbb{R}$ is itself a Kakutani space, with $x \vee y = \max\{x,y\}$, $x \wedge y = \min\{x,y\}$ and $e = 1$. It can be illuminating to use the lattice notations also in this case, even though their use is easily avoided.

As mentioned in Section 3, the vector space $V = C(K)$ of all continuous real-valued functions on a compact Hausdorff space K is a Kakutani space. In fact V is complete with respect to its norm, since $\mathbb{R}$ is complete and the limit of a uniformly convergent sequence of continuous functions is again a continuous function. The Krein–Kakutani theorem says, in effect, that every complete Kakutani space is of this type.

Let V be a Kakutani space and let V' be the vector space of all bounded linear functionals on V. We will show that V' can also be given the structure of a vector lattice.

If $f,g \in V'$ we define $f \le g$ if $f(x) \le g(x)$ for every $x \in V_+$. This evidently defines a partial order on V' with the property (i) in the definition of a vector lattice. It remains to show that any two elements of V' have a least upper bound.

For any $f,g \in V'$ and any $x \in V_+$, put

$$h(x) = \sup\{f(y) + g(x-y) : 0 \le y \le x\}.$$

Then

$$f(x) \le h(x), \ g(x) \le h(x) \ \text{ for every } x \in V_+.$$

Furthermore, if $l \in V'$ is a linear functional such that

$$f(x) \le l(x), g(x) \le l(x) \ \text{ for every } x \in V_+,$$

then

$$f(y) + g(x-y) \le l(y) + l(x-y) = l(x) \ \text{ for } 0 \le y \le x,$$

and hence

$$h(x) \le l(x) \ \text{ for every } x \in V_+.$$

It is clear from the definition of h that $h(\lambda x) = \lambda h(x)$ for every $x \in V_+$ and every $\lambda \in \mathbb{R}_+$, and $|h(x)| \le (\|f\| + \|g\|)\|x\|$. We will show that also

$$h(x_1 + x_2) = h(x_1) + h(x_2) \ \text{ for all } x_1,x_2 \in V_+.$$

If $0 \le y_1 \le x_1$ and $0 \le y_2 \le x_2$, then $y = y_1 + y_2$ satisfies $0 \le y \le x_1 + x_2$. Conversely, if $0 \le y \le x_1 + x_2$ and if we put $y_1 = x_1 \wedge y$, $y_2 = y - y_1$, then $y_1 + y_2 = y$, $0 \le y_1 \le x_1$ and

$$0 \le y_2 = y + (-x_1 \vee -y) = (y - x_1) \vee 0 \le (x_1 + x_2 - x_1) \vee 0 = x_2.$$

Hence

$$\begin{aligned} h(x_1+x_2) &= \sup \{f(y_1) + f(y_2) + g(x_1-y_1) + g(x_2-y_2) : 0 \le y_1 \le x_1, 0 \le y_2 \le x_2\} \\ &= h(x_1) + h(x_2). \end{aligned}$$

We show next that there exists a linear functional $k: V \to \mathbb{R}$ such that $k(x) = h(x)$ for every $x \in V_+$. Every $x \in V$ admits the representation $x = x_+ - x_-$, where $x_+, x_- \in V_+$. In general, if $x = y - z$, where $y,z \in V_+$, we define

$$k(x) = h(y) - h(z).$$

This definition does not depend on the particular representation of x as the difference of two elements of V_+. For if $x = y' - z'$, where $y',z' \in V_+$, then from $y + z' = y' + z$ we obtain

$$h(y) + h(z') = h(y + z') = h(y' + z) = h(y') + h(z)$$

and hence

$$h(y) - h(z) = h(y') - h(z').$$

It may be verified similarly that k is linear,

$$k(x + x') = k(x) + k(x'), \quad k(-x) = -k(x), \quad k(\lambda x) = \lambda k(x) \text{ for } \lambda \in \mathbb{R}_+,$$

and thus is the required least upper bound of f and g.

Because a Kakutani space V has more structure than an arbitrary normed vector space, the supremum norm on the dual space V',

$$\|f\| = \sup \{|f(x)| : \|x\| \le 1\},$$

also has additional properties. These properties will now be investigated.

First, let $f: V \to \mathbb{R}$ be any linear functional such that $f(y) \ge 0$ for every $y \ge 0$. If $x \in V$ and $\|x\| = 1$ then, for any $\varepsilon > 0$, $-(1 + \varepsilon)e \le x \le (1 + \varepsilon)e$ and hence

$$-(1 + \varepsilon)f(e) \le f(x) \le (1 + \varepsilon)f(e).$$

Consequently $|f(x)| \le f(e)$. Thus $f \in V'$ and $\|f\| = f(e)$.

In agreement with the notation already introduced for V, put

$$V'_+ = \{f \in V' : f(y) \ge 0 \text{ for every } y \in V_+\}.$$

If $f \in V'_+$ then $\|f\| = f(e)$, by what we have just proved, and hence

(N5)′ $\|f+g\| = \|f\| + \|g\|$ *if* $f,g \in V'_+$.

This property replaces the property **(N5)** in V. It will now be shown that the property **(N4)** continues to hold in V', i.e.

(N4)′ $\|f\| = \|f^+ + f^-\|$ *for every* $f \in V'$.

We have

$$\begin{aligned}\|f^+ + f^-\| &= f^+(e) + f^-(e)\\ &= \sup\{f(y): 0 \le y \le e\} + \sup\{-f(z): 0 \le z \le e\}\\ &= \sup\{f(y-z): 0 \le y,z \le e\}.\end{aligned}$$

If we put $w = y - z$, then

$$w^+ - w^- = w, \quad w^+ + w^- = w \vee (-w).$$

Hence

$$\|f^+ + f^-\| = \sup\{f(w): w^+ + w^- \le e\}$$

or, since $\|w^+ + w^-\| = \|w\|$,

$$\|f^+ + f^-\| = \sup\{f(w): \|w\| \le 1\} = \|f\|.$$

Now put

$$C = \{f \in V'_+ : \|f\| = 1\}.$$

As we have seen, we can also write

$$C = \{f \in V'_+ : f(e) = 1\}.$$

It follows that C is a convex subset of the unit sphere in V' and that C is closed in the weak* topology. Hence, by the Banach–Alaoglu theorem, C is weak* compact. The extreme points of the convex set C may be characterized in the following way:

(LH) *A linear functional* $f \in C$ *is an extreme point of* C *if and only if it is a lattice homomorphism*, i.e.

$$f(x \vee y) = f(x) \vee f(y), \quad f(x \wedge y) = f(x) \wedge f(y) \text{ for all } x,y \in V.$$

Proof Suppose first that f is a lattice homomorphism. If f is not an extreme point of C, then $f = \theta f_1 + (1-\theta)f_2$ for some $f_1, f_2 \in C$ with $f_1 \ne f_2$ and some $\theta \in (0,1)$.

If $f(x^+) = 0$ for some $x \in V$, then $f_1(x^+) = f_2(x^+) = 0$, and similarly if $f(x^-) = 0$ then $f_1(x^-) = f_2(x^-) = 0$. If $f(x) = 0$ for some $x \in V$, then $f(x^+) = f(x^-) = 0$, since $f(0) = 0$ and f is a lattice homomorphism. Since $f_i(x) = f_i(x^+) - f_i(x^-)$ $(i = 1;2)$, it follows that $f(x) = 0$ implies $f_1(x) = f_2(x) = 0$.

The set $H = \{z \in V: f(z) = 0\}$ is a hyperplane of V which passes through the origin. We have shown that $f_1(z) = f_2(z) = 0$ for every $z \in H$. Since $f_1(e) = f_2(e) = 1$ and every $x \in V$ has the form $x = z + \alpha e$ for some $z \in H$, $\alpha \in \mathbb{R}$, it follows that $f_1(x) = \alpha = f_2(x)$. Thus $f_1 = f_2$, which is a contradiction.

Suppose next that f is an extreme point of C. Assume first that $f(x^+) \wedge f(x^-) = 0$ for every $x \in V$. Since $x = x^+ - x^-$, we have $f(x) = f(x^+) - f(x^-)$. Since $f(x^+) \geq 0, f(x^-) \geq 0$ and min $\{f(x^+), f(x^-)\} = 0$, by hypothesis, it follows that $f(x^+) = f(x)^+, f(x^-) = f(x)^-$. From $x \vee y = y + (x - y)^+$ we now obtain

$$\begin{aligned} f(x \vee y) &= f(y) + f((x-y)^+) \\ &= f(y) + (f(x-y))^+ = f(x) \vee f(y), \end{aligned}$$

and similarly $f(x \wedge y) = f(x) \wedge f(y)$. Thus f is a lattice homomorphism.

Hence to complete the proof we will assume that $f(z^+) \wedge f(z^-) \neq 0$ for some $z \in V$ and derive a contradiction. Evidently z^+, z^- are non-zero and are not scalar multiples of one another. Since $z^+ \wedge z^- = 0$, it follows that $\alpha z^+ + \beta z^- \geq 0$ if and only if $\alpha \geq 0, \beta \geq 0$. Hence $(\alpha z^+ + \beta z^-)^+ = \alpha^+ z^+ + \beta^+ z^-$ for all $\alpha, \beta \in \mathbb{R}$.

Define $p: V \to \mathbb{R}$ by $p(x) = f(x^+)$. Then $p(x) \geq 0$ for every $x \in V$ and $p(x) = 0$ if $x \leq 0$. Moreover p is sublinear. Let Π be the plane determined by $0, z^+, z^-$ and define a bounded linear functional $f_0: \Pi \to \mathbb{R}$ by

$$f_0(\alpha z^+ + \beta z^-) = \alpha f(z^+) \text{ for all } \alpha, \beta \in \mathbb{R}.$$

Since $(\alpha z^+ + \beta z^-)^+ = \alpha^+ z^+ + \beta^+ z^-$, we have

$$\begin{aligned} p(\alpha z^+ + \beta z^-) &= f(\alpha^+ z^+ + \beta^+ z^-) \\ &= \alpha^+ f(z^+) + \beta^+ f(z^-) \geq \alpha f(z^+). \end{aligned}$$

Thus $f_0(x) \leq p(x)$ for all $x \in \Pi$.

It follows from the Hahn–Banach theorem (HB) that there exists a bounded linear functional $f_1: V \to \mathbb{R}$ such that $f_1(x) \leq p(x)$ for all $x \in V$ and $f_1(x) = f_0(x)$ for $x \in \Pi$. Put $f_2 = f - f_1$. If $x \geq 0$ then $f_1(x) \geq 0$, since $-f_1(x) = f_1(-x) \leq p(-x) = 0$, and $f_2(x) \geq 0$, since $f(x) = p(x) \geq f_1(x)$. Evidently also $f_1(z^+) = f(z^+) \neq 0$ and $f_2(z^-) = f(z^-) \neq 0$. Thus if we put $\lambda_i = f_i(e)$ then $\lambda_i > 0$ $(i = 1,2)$ and $\lambda_1 + \lambda_2 = 1$.

Since $f = \lambda_1(\lambda_1^{-1}f_1) + \lambda_2(\lambda_2^{-1}f_2)$, where $\lambda_1^{-1}f_1, \lambda_2^{-1}f_2 \in C$, and since f is an extreme point of C, this is the required contradiction. □

Finally we require the lattice version of the Stone–Weierstrass theorem:

(SW) *Let K be a compact Hausdorff space containing more than one point, let* $\mathrm{C}(K)$ *be the Kakutani space of all continuous real-valued functions on K, and let $\mathfrak{L}$ be a vector sublattice of* $\mathrm{C}(K)$, *i.e. $\mathfrak{L}$ is a subset of* $\mathrm{C}(K)$ *such that if $f,g \in \mathfrak{L}$ and $\alpha,\beta \in \mathbb{R}$ then $\alpha f + \beta g \in \mathfrak{L}$ and $f \vee g \in \mathfrak{L}$.*

Suppose also that $\mathfrak{L}$ contains the constant function 1 *and that $\mathfrak{L}$ separates the points of K, i.e. if $s,t \in K$ and $s \neq t$ then there exists $f \in \mathfrak{L}$ such that $f(s) \neq f(t)$.*

Then $\mathfrak{L}$ is dense in $\mathrm{C}(K)$, *i.e. for any $h \in \mathrm{C}(K)$ and any $\varepsilon > 0$, there exists $f \in \mathfrak{L}$ such that $\|f - h\| < \varepsilon$.*

Proof We first observe that, for any distinct points $s,t \in K$ and any $\alpha,\beta \in \mathbb{R}$, there exists $f \in \mathfrak{L}$ such that $f(s) = \alpha$, $f(t) = \beta$. Indeed if $g \in \mathfrak{L}$ is such that $g(s) \neq g(t)$, we can take

$$f = [g(s) - g(t)]^{-1}\{(\alpha - \beta)g + \beta g(s) - \alpha g(t)\}.$$

Let $h \in \mathrm{C}(K)$ and, for any distinct points $s,t \in K$, let $f_{st} \in \mathfrak{L}$ be such that $f_{st}(s) = h(s)$ and $f_{st}(t) = h(t)$. For given $\varepsilon > 0$, define open subsets O_{st} of K by

$$O_{st} = \{u \in K : f_{st}(u) > h(u) - \varepsilon\}.$$

Then, for each $t \in K$, $K = \bigcup_{s \in K} O_{st}$, since $s \in O_{st}$. Since K is compact, there is a finite subcover:

$$K = O_{s_1 t} \cup \ldots \cup O_{s_m t}.$$

If $f_t = f_{s_1 t} \vee \ldots \vee f_{s_m t}$, then $f_t \in \mathfrak{L}$, $f_t(t) = h(t)$ and $f_t(u) > h(u) - \varepsilon$ for all $u \in K$. Now define an open subset O_t of K by

$$O_t = \{u \in K : f_t(u) < h(u) + \varepsilon\}.$$

Then $K = \bigcup_{t \in K} O_t$, since $t \in O_t$. Let

$$K = O_{t_1} \cup \ldots \cup O_{t_n}$$

be a finite subcover. If we put $f = f_{t_1} \wedge \ldots \wedge f_{t_n}$, then $f \in \mathfrak{L}$ and, for every $u \in K$,

$$h(u) - \varepsilon < f(u) < h(u) + \varepsilon. \quad \square$$

We are now in a position to prove the Krein–Kakutani theorem:

(KK) *If V is a Kakutani space, then there is a linear isometry and lattice isomorphism of V onto a dense subset of the Kakutani space $C(K)$ of all continuous real-valued functions on some compact Hausdorff space K.*

Proof Let

$$C = \{f \in V'_+ : \|f\| = 1\} = \{f \in V'_+ : f(e) = 1\},$$

and let K be the set of all extreme points of C. The theorem will be established in a very explicit manner by proving

(i) K is a compact Hausdorff space with the weak* topology on V',
(ii) for each $x \in V$, the map $J_x: K \to \mathbb{R}$ defined by $J_x(f) = f(x)$ is a continuous function on K,
(iii) the map $J: V \to C(K)$ defined by $J(x) = J_x$ has the properties

$$J(\alpha x + \beta y) = \alpha J(x) + \beta J(y) \text{ for all } x,y \in V \text{ and all } \alpha,\beta \in \mathbb{R},$$
$$\|J(x)\| = \|x\| \text{ for all } x \in V,$$
$$J(x \vee y) = J(x) \vee J(y), \quad J(x \wedge y) = J(x) \wedge J(y) \text{ for all } x,y \in V,$$

(iv) $J(V)$ is dense in $C(K)$ and $J(e)$ is the constant function 1.

We have already seen that the weak* topology on V' is a Hausdorff topology, that C is a compact convex subset of V', and that K is the set of all $f \in C$ which are lattice homomorphisms. It follows that K is closed, and hence compact. Since the map $J'_x: V' \to \mathbb{R}$ defined by $J'_x(f) = f(x)$ is a bounded linear functional, its restriction J_x to K is certainly continuous and

$$J_{\alpha x+\beta y} = \alpha J_x + \beta J_y.$$

Moreover, since every $f \in K$ is a lattice homomorphism,

$$J_{x\vee y} = J_x \vee J_y, \quad J_{x\wedge y} = J_x \wedge J_y.$$

Thus $J(V)$ is a vector sublattice of $C(K)$. Furthermore J_e is the constant function 1, and if f,g are distinct elements of K then $J_x(f) \neq J_x(g)$ for some $x \in V$. Hence, by (SW), $J(V)$ is dense in $C(K)$.

It remains to show that $\|J(x)\| = \|x\|$ for every $x \in V$. In fact we need only verify this for every $x \in V_+$. For if $x^* = x^+ + x^- = x \vee (-x)$, then $\|x^*\| = \|x\|$ and, since every $f \in K$ is a lattice homomorphism, $f(x^*) = \max\{f(x), f(-x)\} = |f(x)|$; hence

$$\|J(x)\| = \sup\{|f(x)| : f \in K\} = \sup\{f(x^*) : f \in K\} = \|J(x^*)\|.$$

Suppose now that $x \geq 0$. Since

$$\|x\| = \sup\{f(x) : f \in V', \|f\| \leq 1\},$$

it follows from (KM) that

$$\|x\| = \sup\{f(x) : f \in K'\},$$

where K' is the set of extreme points of the unit ball $U' = \{f \in V', \|f\| \leq 1\}$. It is evident that if $f \in K'$, then $\|f\| = 1$ and hence, by **(N4)′–(N5)′**,

$$\|f^+\| + \|f^-\| = \|f^+ + f^-\| = \|f\| = 1.$$

It follows that either $f^+ = 0$ or $f^- = 0$, since otherwise

$$f = \|f^+\| \, (f^+/\|f^+\|) + \|f^-\| \, (-f^-/\|f^-\|)$$

would not be an extreme point of U'. Since $x \geq 0$, in evaluating $\|x\|$ we can restrict attention to those $f \in K'$ for which $f = f^+$, and these f are in K. Hence J is an isometry:

$$\|x\| = \sup\{f(x) : f \in K\} = \|J(x)\|. \quad \square$$

5 NOTES

Hausdorff (1927) defined the 'Hausdorff' metric and proved Proposition 3. Blaschke (1916) had already proved Proposition 9.

Rådström (1952) proved Proposition 4 and essentially showed that $\mathscr{C}(X)$ could be embedded in a normed vector space. The lattice properties were added by Pinsker (1966). Schmidt (1986) makes the connection with the Krein–Kakutani theorem and gives applications of Theorem 10 to the theory of random sets and interval mathematics.

If X is complete in Theorem 10, then $\mathscr{C}'$ is also complete, by Proposition 8. However, Debreu (1967) shows that $\mathscr{L}$ is not complete even in the case $X = \mathbb{R}^2$. It should be noted also that if $\mathscr{C}$ is regarded as actually embedded in $\mathscr{L}$, then one must distinguish between $-A$ and $(-1)A$. In fact the two are distinct for any $A \in \mathscr{C}$ with $|A| > 1$.

The Krein–Kakutani theorem is usually formulated for complete Kakutani spaces, or '*AM-spaces with unit*' in the official terminology. It was independently proved by M.G. and S.G. Krein, and by Kakutani, in the years 1940–1941. The theorem is discussed in a more general context in the books of Day (1973), Schaefer (1974) and Meyer-Nieberg (1991). Tychonoff's theorem is proved in Hewitt and Stromberg (1975), for example.

References

J. Bair and R. Fourneau (1975/80), *Etude géométrique des espaces vectoriels* (2 vols.), Lecture Notes in Mathematics **489** and **802**, Springer-Verlag, Berlin.

M.L. Balinski (1961), On the graph structure of convex polyhedra in n-space, *Pacific J. Math.* **11**, 431–434.

I. Bárány (1982), A generalization of Carathéodory's theorem, *Discrete Math.* **40**, 141–152.

I. Bárány and D.G. Larman (1992), A colored version of Tverberg's theorem, *J. London Math. Soc.* **45**, 314–320.

M.K. Bennett and G. Birkhoff (1985), Convexity lattices, *Algebra Universalis* **20**, 1–26.

W. Blaschke (1916), *Kreis und Kugel*, Veit, Leipzig. [2nd ed., de Gruyter, Berlin, 1956]

V.G. Boltyanskii (1975), The method of tents in the theory of extremal problems, *Russian Math. Surveys* **30**, no. 3, 1–54.

T. Bonnesen und W. Fenchel (1934), *Theorie der konvexen Körper*, Springer-Verlag, Berlin. [Reprinted, Chelsea, New York, 1948]

N. Bourbaki (1987), *Topological vector spaces*, Chapters 1–5, Springer-Verlag, Berlin.

A. Brøndsted (1983), *An introduction to convex polytopes*, Springer-Verlag, New York.

H. Brunn (1913), Über Kerneigebiete, *Math. Ann.* **73**, 436–440.

H. Busemann (1955), *The geometry of geodesics*, Academic Press, New York.

J.R. Calder (1971), Some elementary properties of interval convexities, *J. London Math. Soc.* (2) **3**, 422–428.

C. Carathéodory (1911), Über den Variabilitätsbereich der Fourierschen Konstanten von positiven harmonischen Funktionen, *Rend. Circ. Mat. Palermo* **32**, 193–217. [Reprinted in *Gesammelte mathematische Schriften*, Band 3, pp. 78–110, Beck, München, 1955]

W.A. Coppel (1995), A theory of polytopes, *Bull. Austral. Math. Soc.* **52**, 1–24.

H.S.M. Coxeter (1989), *Introduction to geometry*, 2nd ed., reprinted by Wiley, New York.

L. Danzer, B. Grünbaum and V. Klee (1963), Helly's theorem and its relatives, *Convexity* (ed. V. Klee), pp. 101–180, Proc. Symp. Pure Math. **7**, Amer. Math. Soc., Providence, R.I.

M.M. Day (1973), *Normed linear spaces*, 3rd ed., Springer-Verlag, Berlin.

G. Debreu (1967), Integration of correspondences, *Proc. Fifth Berkeley Symp. Math. Statist. Probability* (ed. L.M. Le Cam and J. Neyman), Vol. II, Part 1, pp. 351–372, University of California Press, Berkeley.

J.-P. Doignon (1973), Convexity in cristallographic lattices, *J. Geom.* **3**, 71–85.

J.-P. Doignon (1976), Caractérisations d'espaces de Pasch–Peano, *Acad. Roy. Belg. Bull. Cl. Sci.* (5) **62**, 679–699.

P. Duchet (1987), Convexity in combinatorial structures, *Rend. Circ. Mat. Palermo* (2) *Suppl. No.* **14**, 261–293.

J. Eckhoff (1979), Radon's theorem revisited, *Contributions to geometry* (ed. J. Tölke and J.M. Wills), pp. 164–185, Birkhäuser, Basel.

J. Eckhoff (1993), Helly, Radon, and Carathéodory type theorems, *Handbook of convex geometry* (ed. P.M. Gruber and J.M. Wills), Volume A, pp. 389–448, North-Holland, Amsterdam.

P.H. Edelman and R.E. Jamison (1985), The theory of convex geometries, *Geom. Dedicata* **19**, 247–270.

J.W. Ellis (1952), A general set-separation theorem, *Duke Math. J.* **19**, 417–421.

M. Erné (1984), *Chains, directed sets and continuity*, Institut für Mathematik, Universität Hannover, Preprint Nr. 175.

Euclid (1956), *The thirteen books of Euclid's elements*, English translation by T.L. Heath, 2nd ed., reprinted in 3 vols. by Dover, New York.

R. Frank (1992), Ein lokaler Fundamentalsatz für Projektionen, *Geom. Dedicata* **44**, 53–66.

H. Grassmann (1844), *Die lineale Ausdehnungslehre*, Leipzig. [Reprinted in *Gesammelte Werke* I_1, Chelsea, New York, 1969]

B. Grünbaum (1967), *Convex polytopes*, Interscience, New York.

R. Hammer (1977), Beziehungen zwischen den Sätzen von Radon, Helly und Caratheodory bei axiomatischen Konvexitäten, *Abh. Math. Sem. Univ. Hamburg* **46**, 3–24.

F. Hausdorff (1927), *Mengenlehre*, 2nd ed., de Gruyter, Berlin. [English translation of 3rd ed., *Set theory*, Chelsea, New York, 1962]

E. Helly (1923), Über Mengen konvexer Körper mit gemeinschaftlichen Punkten, *Jahresber. Deutsch. Math.-Verein.* **32**, 175–176.

G. Hessenberg (1905), Beweis des Desarguesschen Satzes aus dem Pascalschen, *Math. Ann.* **61**, 161–172.

E. Hewitt and K. Stromberg (1975), *Real and abstract analysis*, 3rd printing, Springer-Verlag, New York.

A. Heyting (1980), *Axiomatic projective geometry*, 2nd ed., North-Holland, Amsterdam.

D. Hilbert (1899), *Grundlagen der Geometrie*, Berlin. [12th ed., Teubner, Stuttgart, 1977; English translation of 10th ed., *Foundations of geometry*, Open Court, LaSalle, Ill., 1971]

R.B. Holmes (1975), *Geometric functional analysis and its applications*, Springer-Verlag, New York.

R.E. Jamison (1974), *A general theory of convexity*, Dissertation, University of Washington, Seattle.

R.E. Jamison-Waldner (1982), A perspective on abstract convexity: classifying alignments by varieties, *Convexity and related combinatorial geometry* (ed. D.C. Kay and M. Breen), pp. 113–150, Dekker, New York.

B. Jónsson (1959), Lattice-theoretic approach to projective and affine geometry, *The axiomatic method* (ed. L. Henkin *et al.*), pp. 188–203, North-Holland, Amsterdam.

V. Klee (1959), Some characterizations of convex polyhedra, *Acta Math.* **102**, 79–107.

F. Klein (1873), Ueber die sogenannte Nicht-Euklidische Geometrie (Zweiter Aufsatz), *Math. Ann.* **6**, 112–145.

B. Korte and L. Lovász (1984), Shelling structures, convexity and a happy end, *Graph theory and combinatorics* (ed. B. Bollobás), pp. 219–232, Academic Press, London.

Z. Kovijanić (1994), A proof of Bárány's theorem, *Publ. Inst. Math. (Beograd) (N.S.)* **55** (69), 47–50.

S. Lang (1984), *Algebra*, 2nd ed., Addison-Wesley, Reading, Mass.

M. Lassak (1986), A general notion of extreme subset, *Compositio Math.* **57**, 61–72.

K. Leichtweiss (1980), *Konvexe Mengen*, Springer-Verlag, Berlin.

H. Lenz (1992), Konvexität in Anordnungsräumen, *Abh. Math. Sem. Univ. Hamburg* **62**, 255–285.

F.W. Levi (1951), On Helly's theorem and the axioms of convexity, *J. Indian Math. Soc.* **15**, 65–76.

F. Maeda and S. Maeda (1970), *Theory of symmetric lattices*, Springer-Verlag, Berlin.

K. Menger (1936), New foundations of projective and affine geometry, *Ann. of Math.* **37**, 456–482.

P. Meyer-Nieberg (1991), *Banach lattices*, Springer-Verlag, Berlin.

F.R. Moulton (1902), A simple non-desarguesian plane geometry, *Trans. Amer. Math. Soc.* **3**, 192–195.

J.G. Oxley (1992), *Matroid theory*, Oxford University Press, New York.

M. Pasch (1882), *Vorlesungen über neuere Geometrie*, Teubner, Leipzig. [2nd ed. with appendix by M. Dehn, reprinted by Springer-Verlag, Berlin, 1976]

G. Peano (1889), *I principii di geometria logicamente esposti*, Fratelli Bocca, Torino. [Reprinted in *Opere scelte*, Vol. II, pp. 56–91, Cremonese, Roma, 1958.]

G. Pickert (1981), Projectivities in projective planes, *Geometry – von Staudt's point of view*, (ed. P. Plaumann and K. Strambach), pp. 1–49, Reidel, Dordrecht, Netherlands.

A.G. Pinsker (1966), The space of convex sets of a locally convex space (Russian), *Leningrad Inzh.-Ekonom. Inst. Trudy Vyp.* **63**, 13–17.

A.V. Pogorelov (1991), On a theorem of Beltrami, *Soviet Math. Dokl.* **43**, 83–85.

W. Prenowitz (1961), A contemporary approach to classical geometry, *Amer. Math. Monthly* **68**, no. 1, part II, 67 pp.

W. Prenowitz and J. Jantosciak (1979), *Join geometries*, Springer-Verlag, New York.

S. Priess-Crampe (1983), *Angeordnete Strukturen: Gruppen, Körper, projektive Ebenen*, Springer-Verlag, Berlin.

J. Radon (1921), Mengen konvexer Körper, die einen gemeinsamen Punkt enthalten, *Math. Ann.* **83**, 113–115. [Reprinted in *Collected works*, Vol. 1, pp. 367–369, Birkhäuser, Basel, 1987]

H. Rådström (1952), An embedding theorem for spaces of convex sets, *Proc. Amer. Math. Soc.* **3**, 165–169.

W. Rinow (1961), *Die innere Geometrie der metrischen Räume*, Springer-Verlag, Berlin.

G.S. Rubinštein (1964), Theorems on the separation of convex sets (Russian), *Sibirsk. Mat. Zh.* **5**, 1098–1124.

H.R. Salzmann (1967), Topological planes, *Adv. in Math.* **2**, 1–60.

K.S. Sarkaria (1992), Tverberg's theorem via number fields, *Israel J. Math.* **79**, 317–320.

H.H. Schaefer (1974), *Banach lattices and positive operators*, Springer-Verlag, Berlin.

K.D. Schmidt (1986), Embedding theorems for classes of convex sets, *Acta Appl. Math.* **5**, 209–237.

R. Schneider (1993), *Convex bodies: the Brunn–Minkowski theory*, Cambridge University Press.

A. Schrijver (1989), *Theory of linear and integer programming*, Wiley, Chichester.

F. Schur (1909), *Grundlagen der Geometrie*, Teubner, Leipzig.

B. Segre (1956), Plans graphiques algébriques réels non Desarguésiens et correspondances crémoniennes topologiques, *Rev. Roumaine Math. Pures Appl.* **1**, no. 3, 35–50.

A. Seidenberg (1976), Pappus implies Desargues, *Amer. Math. Monthly* **83**, 190–192.

G. Sierksma (1984), Extending a convexity space to an aligned space, *Nederl. Akad. Wetensch. Indag. Math.* **46**, 429–435.

D.E. Smith (1959), *A source book in mathematics*, Dover, New York.

V.P. Soltan (1984), *Introduction to the axiomatic theory of convexity* (Russian), Shiintsa, Kishinev.

E. Sperner (1938), Zur Begründung der Geometrie im begrenzten Ebenenstück, *Schriften der Königsberg Gelehrten Gesellschaft* **6**, 121–143.

M. Spivak (1975), *A comprehensive introduction to differential geometry*, Vol. IV, Publish or Perish, Boston, Mass.

L.A. Steen and J.A. Seebach Jr (1978), *Counterexamples in topology*, 2nd ed., Springer-Verlag, New York.

J. Stoer and C. Witzgall (1970), *Convexity and optimization in finite dimensions* I, Springer-Verlag, Berlin.

L.W. Szczerba (1972), A paradoxical model of Euclidean affine geometry, *Bull. Acad. Polon. Sci. Sér. Sci. Math. Astron. Phys.* **20**, 845–851.

L.W. Szczerba and A. Tarski (1979), Metamathematical discussion of some affine geometries, *Fund. Math.* **104**, 155–192.

F.A. Toranzos (1967), Radial functions of convex and star-shaped bodies, *Amer. Math. Monthly* **74**, 278–280.

S.N. Tschernikow (1971), *Lineare Ungleichungen*, VEB Deutscher Verlag der Wissenschaften, Berlin. [Russian original, 1968]

H. Tverberg (1966), A generalization of Radon's theorem, *J. London Math. Soc.* **41**, 123–128.

H. Tverberg and S. Vrećica (1993), On generalizations of Radon's theorem and the ham sandwich theorem, *European J. Combin.* **14**, 259–264.

F.A. Valentine (1964), *Convex sets*, McGraw-Hill, New York.

M.L.J. van de Vel (1993), *Theory of convex structures*, North-Holland, Amsterdam.

O. Veblen (1904), A system of axioms for geometry, *Trans. Amer. Math. Soc.* **5**, 343–384.

O. Veblen and J.W. Young (1910/18), *Projective geometry* (2 vols.), Ginn, Boston, Mass. [Reprinted by Blaisdell, New York, 1965]

K.G.C. von Staudt (1856/7), *Beiträge zur Geometrie der Lage* (2 vols.), Nürnberg.

C.T.C. Wall (1972), *A geometric introduction to topology*, Addison-Wesley, Reading, Mass.

D.J.A. Welsh (1976), *Matroid theory*, Academic Press, London.

N. White (ed.) (1992), *Matroid applications*, Cambridge University Press.

G.M. Ziegler (1995), *Lectures on polytopes*, Springer-Verlag, New York.

Notations

Axioms

Propositions

Examples

Index